MAINTAINING LONGCASE CLOCKS

An Owner's Guide to Maintenance, Restoration and Conservation

MAINTAINING LONGCASE CLOCKS

An Owner's Guide to Maintenance, Restoration and Conservation

Nigel Barnes and Austin Jordan

THE CROWOOD PRESS

First published in 2013 by
The Crowood Press Ltd
Ramsbury, Marlborough
Wiltshire SN8 2HR

www.crowood.com

© Nigel Barnes and Austin Jordan 2013

All rights reserved. No part of this publication may be reproduced or transmitted in any form or by any means, electronic or mechanical, including photocopy, recording, or any information storage and retrieval system, without permission in writing from the publishers.

British Library Cataloguing-in-Publication Data
A catalogue record for this book is available from the British Library.

ISBN 978 1 84797 521 8

Disclaimer
The practical workshop procedures and the tools and equipment used by clock-repairers and conservationists are potentially dangerous. All equipment and tools employed in clock repair and conservation work should be used in strict accordance with both current health and safety regulations and the manufacturer's instructions.

    The author and the publisher do not accept any responsibility in any manner whatsoever for any error or omission, or any loss, damage, injury, adverse outcome, or liability of any kind incurred as a result of the use of any of the information contained in this book, or reliance upon it. If in doubt about any aspect of the maintenance, restoration, conservation, or repair of longcase clocks, readers are advised to seek professional advice.

Photographs by Nigel Barnes except where otherwise stated.

Typeset by Jean Cussons Typesetting, Diss, Norfolk

Printed and bound in India by Replika Press Pvt Ltd

Contents

	Preface	6
Chapter 1	A Short Introduction to Longcase Clocks	8
Chapter 2	Dating Clocks	40
Chapter 3	Materials, Tools and Equipment	57
Chapter 4	Clock Maintenance	74
Chapter 5	The Movement – Some Simple Workshop Procedures	96
Chapter 6	More Difficult Movement Procedures	114
Chapter 7	The Most Difficult Workshop Procedures	128
Chapter 8	Case-Work – Maintenance and Simple but Effective Repairs	144
Chapter 9	More Difficult Woodwork Procedures	162
	Glossary	*186*
	Bibliography	*190*
	Index	*191*

Preface

From its early beginnings in the thirteenth century, the craft of clockmaking expanded rapidly in the mid-seventeenth century and longcase clocks first appeared at about the time of the restoration of the English monarchy in 1660; Charles II (d. 1685) was an early clock collector.

Since those early days, the emphasis of the many books written on various aspects of horology has shifted towards the appreciation and preservation of valuable horological artefacts.

Times and circumstances change and a few decades ago, clock-repair shops were plentiful and clock-repairers busy, but more recently, with the introduction of electric clocks, the old generation of time-served craftsmen has largely disappeared. That sad loss, compounded with the scarcity of formal training facilities for the next generation of clock-menders, limits the antique clock-owners' options for maintenance, conservation and restoration work.

No book about longcase clocks can ever be complete, but in the evolving environment of more self-reliance, we have provided here a certain amount of background information and guidance for interested connoisseurs and owners of longcase clocks and, of course, aspiring menders. We describe in detail a wide range of repair work that will guide the amateur, or at least explain what is involved. However, this could never be a complete workshop manual for all longcase clocks because, in our combined experience, no two are identical; not only as originally conceived, but also through their individual history of repairs.

Traditionally, the clock-repairer was something of a pragmatist who, when faced with a problem, could engineer a sound repair without necessarily needing to consider the clock as an important historical artefact. In most cases a clock was just a day-to-day domestic appliance and he was more likely to have been primarily concerned with the proper functioning of a clock as a time-keeper, than in preserving the evidence of its place in the history of horology.

Nowadays we might regret that approach but, as with all historical artefacts, their increasing age and scarcity force changes in our perceptions about worth and importance.

Both of us have encountered and dealt with bizarre and often ineffective repairs, and we both keep to the rule of never making value judgements about such work. As Austin's father said, 'You never know about the full circumstances of repair work – you don't know what tools that repairer had and what training, just accept that he was doing his best and work around it and try to do your work in a way that the original maker would approve.'

We also accept that, unlike the professional who needs to do his work to a good standard, the owner of a clock is entitled to do as he pleases with his own property. So from time to time we include gentle reminders, with phrases such as 'historical integrity', that hand-made, longcase clocks of the type we describe were, in general, not made after the mid-nineteenth century: they are invariably old or, in some cases, very old and as a species they are becoming endangered.

Likewise, attitudes to the restoration of clock-cases have developed to recognize the concepts of the conservation of the historically important, as against restoration with a more functional objective.

We do not set any limits to what an enthusiastic amateur can achieve, although we occasionally mention that practice and patience are required at the work-bench, especially for some of the more awkward procedures. The other aspect of longcase clock-repair that we touch on is the necessary broad understanding of the subject gradually acquired over the years by reading and research. That understanding also includes an appreciation of the social and economic development of the British Isles, because it enables an interpretation of the origins and context of any particular clock, which will be helpful in planning an appropriate repair.

Prior to the rise of consumerism and mass production in the early nineteenth century, clockmakers either undertook, or were at least able to undertake, all aspects of the work involved in making a clock movement, and we have attempted to give some, far too brief, guidance along the same principle. So in several instances where we mention that certain replacement parts are available from horological suppliers, we also describe the method of making them from scratch.

We hope that the reader will develop an interest in longcase clocks and build on this and former work to preserve these wonderful horological treasures, not as dead museum exhibits but as living, ticking contrivances that are redolent of the golden age of craftsmanship.

While Nigel describes himself as an enthusiastic collector from a long line of clockmakers, Austin is a formally trained clock-mender (also from a long line of clock-men). The approach taken here to explain about repair procedures and how the various parts work is based partly on individual experiences and partly on joint experiences, including running weekend clock-restoration and conservation courses.

With just a couple of exceptions, the clocks illustrated are from Nigel's own longcase collection, which covers a span of about a hundred and fifty years, from the age of London clocks, around the year 1700, through the age of provincial clocks to the rise of modern society and mass production, which brought a virtual end to hand-made longcase clock production.

This book offers advice about practical workshop procedures and the reader should ensure that any tools or materials are used carefully and in accordance with the maker's recommendations: accidents and injuries are not in keeping with the professionalism that befits a serious clock-worker.

Chapter 1

A Short Introduction to Longcase Clocks

TERMINOLOGY AND THE WORDS: REPAIR, RESTORATION AND CONSERVATION

'Clock-repairer' is a term that is used to describe people who correct wear and defects in clocks and although we use the word 'repair' rather loosely to cover all corrective procedures, 'restoration' is a more appropriate concept for longcase clocks because they are invariably old and could be considered as horological artefacts. In more extreme cases of age, rarity or quality of workmanship, the 'conservation' approach implies stabilization and preservation of important heritage objects, even at the expense of functionality.

For clarity, the difference between the concepts of repair and restoration is explained by defining the terms:

- **Repair** means a purely functional mending process intended to return a clock to working order without any particular consideration of the clock as it was when first made.
- **Restoration** means returning a clock to its supposed original state sympathetically, using the same style of craftsmanship and materials. An element of heritage value is implied.
- **Conservation** – unlike museums, private owners of historical clocks generally prefer to have them working rather than in a state of suspended silence, preserved for all time, so the concept of conservation often presents conflicts and difficulties for both owners and conservators. Essentially, conservation means preserving original parts in their original context or relationship, and if some are missing or beyond use, any replacement parts should only be introduced to make sense of the original, with no attempt to mimic what was once there. Likewise, with case-work, new work should only be introduced into a clock-case in order to stabilize and preserve the structure. New work should be clearly identifiable, with no attempt to make new parts that might be mistaken for original. Conservation work should be carefully recorded with illustrated descriptions of the clock and all the work including evidence of previous work or alterations.

A responsible clock-mender takes care to preserve a horological artefact as far as possible, while keeping it in a good state of repair.

Craftsman's hand and artist's eye. The proportions of this quite small provincial clock, made in Oxfordshire in the 1770s, suggest that, in addition to good clock work, fashionable design was a consideration.

THE BASIC CONCEPT OF A PENDULUM CLOCK

It is a basic law of physics that over small arcs of swing, and gravitational forces being constant, the time taken for a pendulum to swing from one extremity to the opposite is only dependent on its length. A pendulum clock is designed and built around this isochronous behaviour of simple pendulums.

The movement in a longcase (or any mechanical) clock performs two separate but related tasks: it maintains the swing of the pendulum by giving it a little nudge at each swing to make up for natural damping; and, second, it counts the number of swings, which it shows

The length of a pendulum is taken to be the distance from the point of suspension to the centre of gravity (quite close to the centre of the bob). A regulating screw is used to make small adjustments – a 1mm adjustment is equivalent to roughly 43sec per day.

Classic London styles are recognizable in earlier country clocks. This oak-cased clock was made in Yaxley, Suffolk in the mid-1700s by Thomas Henson. It is tempting to conjecture that he is actually the Thomas Hanson who is recorded as apprenticed in London in 1745.

as seconds, minutes and hours of the day. The word 'dial' is derived from the same root as our word 'day', because the clock dial is a representation of the day.

HOW THE TWO MAIN TYPES OF LONGCASE CLOCK EVOLVED

The terms, 'longcase', 'tall case' (in North America) or otherwise 'grandfather' are used to describe floor-standing, weight-driven clocks that have their dials approximately at adult eye-level. A wooden case supports the movement and dial at that convenient height, while providing a vertical space through which the driving weights descend and the approximately one metre-long pendulum swings.

The fact that the pendulum is roughly one metre long is no coincidence: in 1790 the French National Assembly proposed to define the metre as the length of a pendulum with a half-period of one second. (In other words, by that definition, a 1m-long pendulum takes exactly 1sec to swing from one extremity to the other.) In the event, an alternative, geodetic, definition was adopted, so the one-second pendulum is actually 994.16mm or 39.14in.

Since a longcase clock is both functional and decorative, the case is necessarily both structurally sound and aesthetically pleasing. Like all good furniture, a well-designed and made clock-case is the meeting of the craftsman's hand with the artist's eye.

From the end of the seventeenth century and into the early eighteenth century, when longcase clocks became fashionable in London, they were fabulously expensive symbols of status and wealth. However, after about 1730, production of longcase clocks in London began to decline as fashions changed once more, this time in favour of bracket clocks. About the same time, there was a rapid expansion in clockmaking activity in the provinces and from then, the continual interpretation of national fashions, coupled with the development of regional styles, usually make deductions possible about when and where a longcase clock might have been made and the status of the first buyers, because longcase-clock ownership ceased to be the sole prerogative of the very wealthy.

By the 1740s, approximately fifteen or twenty years after the general introduction of the break-arch style of longcase dial, longcase clocks had ceased to be at the forefront of London fashion. Thereafter, with the London makers concentrating on bracket clocks, London longcase clocks made after the mid-1700s are surprisingly uncommon.

However, the period also saw the beginnings of a huge proliferation of country longcase clocks throughout the towns and villages of the British Isles. In many instances, young men who had migrated to London to be apprenticed to good makers, returned with their new skills to their native towns and villages. Of the 70,000 clockmakers listed in Brian Loomes's *Directory of Watchmakers and Clockmakers of the World*, a very high proportion were working in country towns and villages in the hundred years from the mid-eighteenth century to the end of longcase production in the mid-nineteenth century.

In the first-half of the eighteenth century, as longcase clocks increasingly transcended the social strata, large variations emerged, both in their style and complexity, from plain country versions to the most elaborate examples from the golden age of cabinet-making.

Although movements do not vary much, they usually contain subtle clues about approximate age, but dials and cases continued to develop until the effective end of longcase clock production in the mid- to late nineteenth century, when cheaper mass-produced clocks became widely available.

Within all that variation, there are naturally a number of rules about dating longcase clocks based on their movements, dials and cases, even their hands, but there are exceptions and, with longcase clocks, there are a great many exceptions. The one rule for which there seems to be no exception is that there are no two identical longcase clocks, nor even two identical dials.

THE MECHANICAL DIFFERENCES BETWEEN THE TWO TYPES OF LONGCASE MOVEMENT

In broad terms there are two main types of longcase clock movement: first, those having two weights, suspended by cords of cat-gut, each wound onto a drum by a key inserted through holes in the dial; and, second, those having a single weight wound by pulling an endless rope or chain. Two-weight clocks usually have a duration between re-winding of eight days, hence the generic name, 'eight-day' clocks (effectively, weekly winding), while the pull-up type usually, but not always, has a duration between re-winding of not much more than a day, hence the name '30-hour'.

Within these two quite similar and yet quite different basic clock layouts, there are many variations: clocks designed to run as long as a full year between windings; clocks made for extreme accuracy, even to forecast the time of high tide at a local seaport; and clocks designed

An eight-day dial: because eight-day movements are invariably wound by a key, two winding holes are made in the dial.

A 30-hour clock is almost always wound by pulling on the chain to raise the driving weight. Consequently there are no winding holes in the dial and 30-hour clocks only rarely have seconds hands.

to amuse with a collection of tunes sounded on bells or automata depicting rocking ships or people in various activities.

The vast majority of longcase clock movements contain two trains of gears: one driving the hands and the pendulum, the watch or going train; and the other driving the clock or strike train. With the older names, watch and clock, for the two trains (clock from the Latin, 'clocca' – a bell), it is not unusual to find the letters 'C' and 'W' scribed onto the two drums of eight-day clock movements by an early repairer, those two parts being the only ones that are more or less interchangeable.

EARLY BIRDCAGE MOVEMENTS THAT RETAIN MANY OF THE FEATURES OF SEVENTEENTH-CENTURY TABLE CLOCKS

The longcase clock that is recognizable today developed in the second-half of the seventeenth century. While it relied on the development of two of its parts – the one-second pendulum and the anchor-recoil escapement – it is also a direct development from a previous form of clock. Verge escapement, short pendulum, lantern clocks, made from the mid- to late 1600s principally in London, share some features with early London-made and much later English provincial longcase clocks. They are both constructed with the two wheel-trains mounted one behind the other (strike to the back) in an open framework, usually described as a posted or birdcage construction. They each have a bell for the hourly strike directly over the movement and, in common with lantern clocks, the bell of a birdcage longcase movement is struck by a hammer located on the inside of the bell; but, most significantly, the actual layout of the two sets of moving parts is effectively the same.

The two examples below only have a single (hour) hand moving over the usual silvered chapter ring in which the deeply engraved Roman numeral hours are filled with black wax. The ornamental brass-work and overall design are so similar that they might have been made by the same hand. Indeed, the maker of the 30-hour longcase, Benjamin Shuckforth, is known to have made lantern clocks after he returned to his native Norfolk from London, where he served his apprenticeship.

PLATED CONSTRUCTION

Although provincial 30-hour clocks occasionally retained the posted or birdcage layout well into the eighteenth century, by far the more usual configuration of the two wheel-trains is side-by-side between two solid brass plates held apart by (usually) four brass pillars. The side-by-side arrangement is referred to as 'plated' construction.

Plated movements are practically universal from the late seventeenth century for eight-day and longer duration movements but, on the other hand, it is not impossible to find a birdcage movement dating from the painted iron dial period at end of the eighteenth century.

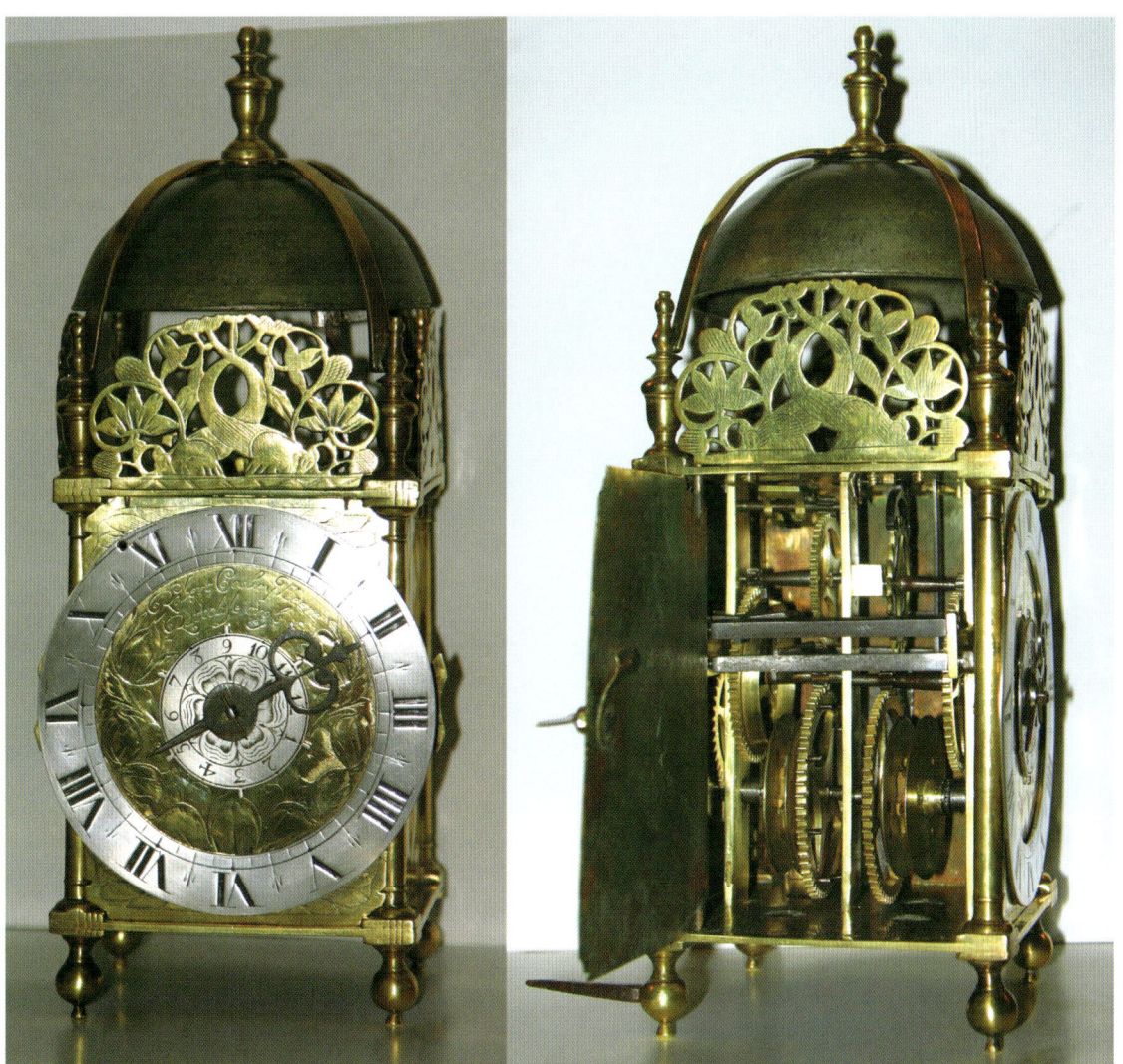

Front and side views of a lantern clock made in London in the 1660s. (Dennis Benson, London Clocks, Worcester)

The dial and hand of an early eighteenth-century birdcage longcase clock. Behind the dial, upper and lower plates support four corner posts and the narrow, central, vertical strip-plates in the same layout as earlier lantern clocks.

rope and the driving force is never relieved from the going side of the movement during winding, so it always drives the movement clockwise. That quirky behaviour of the eight-day clock, which obviously affects the time-keeping, is got around in more accurate or 'better' movements by introducing a subsidiary means of maintaining the power on the clock wheels while it is being wound.

Early plated movements used a system known as 'bolt and shutter', where in raising two spring-loaded plates behind the winding holes, the spring keeps some pressure on the going side train for a few moments while the clock is being wound. Bolt and shutter clocks are recognizable by the plate that blanks off the winding holes and, if that has been removed, by the way the winding squares are recessed. Later, John Harrison, famous for his work on maritime chronometers, developed a simpler maintaining-power device using a spring and ratchet wheel on the great wheel.

Unlike the sudden and universal adoption of plated construction for eight-day clocks from the late seventeenth century, the gradual change from birdcage to plated frames for 30-hour movements seems to have been slower in certain areas and certain makers or groups of makers. The plated movement is intrinsically more rigid, but for reliability and ease of repair work there is little difference between the two types.

One of the idiosyncrasies of normal eight-day clocks is their behaviour during winding, when the driving power is temporarily relieved from the movement. An odd effect can often be seen during winding an eight-day clock that has a seconds hand, when the seconds hand tends to move in an anticlockwise direction. The explanation is that once the driving force is relieved from the movement, rather than the escapement wheel driving the pendulum, the pendulum causes the escapement wheel to rotate, but in reverse. That does not happen with the 30-hour type of clock, where the single weight is suspended by an endless chain or

Typical early plated movement. The maker has made no attempt to smooth the front face of the front-plate, which retains marks from hammering and sand-casting.

A Short Introduction to Longcase Clocks 15

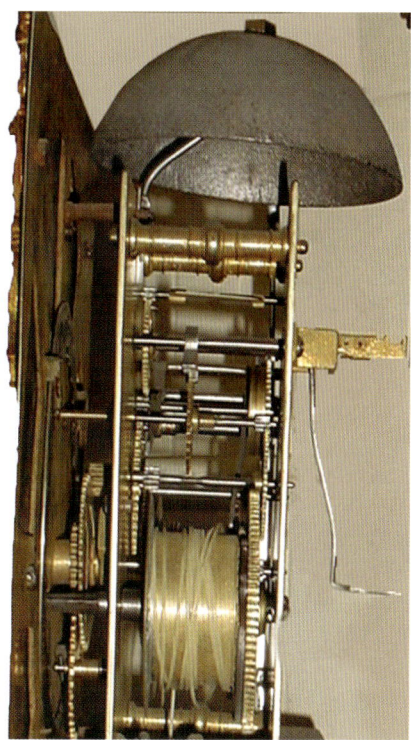

An early provincial clock made in Rochdale about 1700. While the dial is similar in style to contemporary London clocks, the layout of the movement is strikingly different. Although the pillars and wheel-work are typical of 1700, the plates are very tall when compared to London movements.

THE LESS COMMON TYPES OF CLOCK

In general, longcase movements made before the mid-eighteenth century tend to be horologically interesting, often individualistic in layout and, occasionally, even experimental. Although the early London makers are credited with much of the technical innovation in the development of longcase movements, with the introduction of rack-striking, the dead-beat escapement and alternative toothcounts in the wheel-work, developments were also made in the provinces. In 1713, John Harrison made his first longcase clock movement in the North Lincolnshire village of Barrow-upon-Humber, using wood wheels; and advances were made in clockmaking, especially in the north-west.

VARIATIONS IN MOVEMENT DESIGN IN THE FIRST-HALF OF THE EIGHTEENTH CENTURY

Longer Duration Versions

For some reason, single-weight clocks appear not to have been made for longer durations, but long-duration versions of the eight-day clock – one-month, three-month and even one-year durations between winding – were achieved by the introduction of more gearing and heavier weights. The fashion for long-duration clocks in the early part of the eighteenth century is difficult to reconcile at a time when clocks were owned exclusively by the wealthy who had plenty of household staff to wind clocks, and it is reasonable to conclude that there was an element of 'one-upmanship' involved in this otherwise ostensibly futile development of

Ways of arranging the tooth counts for a one-second pendulum – the table shows a few examples for 30-hour, three-wheel clocks

Hour wheel	Pinion of report	Great wheel	Second pinion	Second wheel	Escape pinion	Escape wheel
48	8	60	6	60	6	36
60	10	72	6	60	6	30
72	12	72	6	60	6	30
48	16	90	6	72	6	40
60	15	90	6	72	6	30

horological excellence. It seems that the names of 'good makers' are most frequently associated with longer duration clocks, and it is quite likely that for the clockmaker, a long-duration clock bearing his name would have been a useful marketing feature. As a rule, while the usual eight-day clock has four wheels in the going train, a month-going clock has five wheels and (extremely rare) year-duration clocks have six. In both cases, the plates and pillars are more massive to ensure that the frame remains rigid, despite the heavier weights.

Movements with 'Odd' Pendulum Lengths

There are several ways of arranging the tooth counts for a one-second pendulum and the table above shows a few examples for 30-hour, three-wheel clocks.

However, a surprising number of longcase movements, especially country-made, three-wheel, 30-hour clocks, are made with a pendulum beat that is not exactly one second. That type of clock has no seconds hand, so there would have been no particular reason for the maker to stick rigidly to an exactly one-second pendulum layout. There are a great many possible wheel counts for beats that only approximate to one second, many of which are documented in English 30 Hour Clocks by Jeff Darken and John Hooper.

Movements with 1¼sec Pendulums

A special case of the 'odd' pendulum clocks – the 1¼sec pendulum clock – seems to have

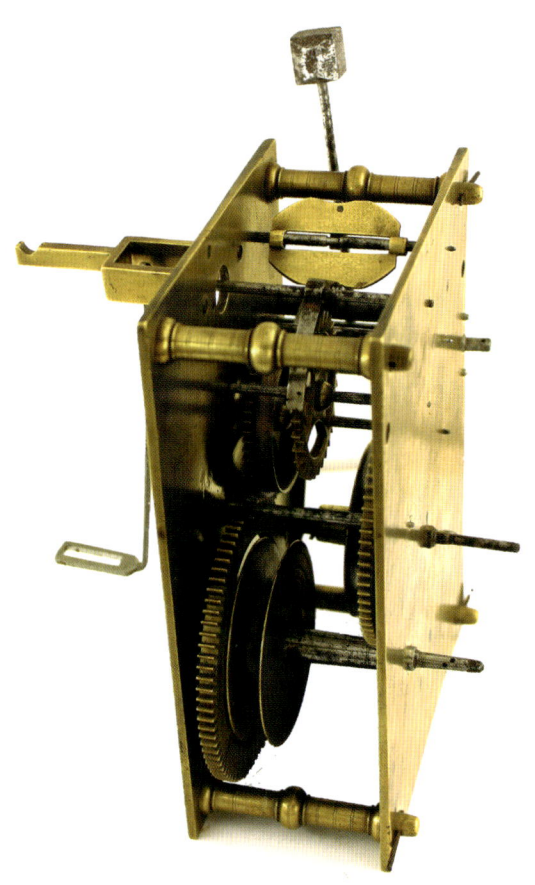

At first sight the movement of this early 30-hour clock is not unusual, but the tooth count reveals an impossible gearing onto the (missing) hour wheel. The clue is in the thirty-four tooth escapement wheel. Recalculating on the basis of a 1¼sec pendulum results in an elegant ratio of 15:14 to drive the motion-work.

A Short Introduction to Longcase Clocks 17

The movement of a Victorian domestic regulator. The characteristically tall deadbeat escapement (with forward-pointing escape wheel teeth) is accompanied by the pawl and ratchet wheel to the front of the main wheel.

disappeared by the end of the seventeenth century. Extant examples are very rare, but they turn up from time to time and can often be identified by the lenticle in the base, and often by wheel counts that cannot be reconciled with 1sec pendulums. Despite the great length of the pendulum at 61½in (1,563mm), the clock cases are of similar height to their one-second counterparts. Only two examples are described by Cesinsky and Webster in *English Domestic Clocks*: one by William Clement, 74in (1,865mm) high and the other by Thomas Tompion, less than 78in (1,981mm) high. Even the Wetherfield collection of 150 or so 'old master' longcases included only four examples and each of those was made by a prestigious London maker.

Precision Movements

Clocks designed for accurate astronomical observations appeared in the late eighteenth century, and precision movements with temperature-compensated pendulums, dead-beat escapements and built-in maintaining-power arrangements, were more widely produced in the early eighteenth century. Usually known as regulators, early examples used the bolt and shutter type of maintaining power, but later, the Harrison spring and ratchet type was the norm.

An accuracy of plus or minus a few seconds a day is quite acceptable for domestic use, but in the days when there was no radio to set the clock by, a very accurate clock was useful as a standard for other clocks in the locality. The temperature expansion of pendulum rods was compensated in more complex pendulums or minimized by using wooden rods, which are far less susceptible to thermal expansion and contraction than metal.

Musical Clocks

More complex chime-work, often with a third train of wheels to drive a set of hammers that struck bells in sequence, first appeared in the late seventeenth century, but a third chime train was an additional feature often adapted into older clocks, especially during the nineteenth century.

Astronomical and Celestial Clocks

Astronomical features, such as moon phase and times of high tides, were often built into clock movements and displayed in the dial arch. Because of the approximately 29½ days of the lunar cycle, the tooth ratios of the wheel-work are usually simplified by using a 118-tooth wheel with two moons, moving one tooth every 12h. A more accurate wheel train is not recorded until that devised in the nineteenth century by Lord Grimthorpe, designer of the Big Ben movement.

At certain times of the year, solar and average time may vary by nearly a quarter of an hour. The elusive connection between solar time and average time is frequently incorporated into 'equation of time' clocks: the usual arrangement of showing the difference is a pointer attached to a cam on a wheel that rotates once in a year.

Four-Wheel or Centre-Wheel 30-Hour Clocks

Although the going train of a 30-hour clock usually comprises three wheels – the great wheel, the inaccurately named third wheel and the escapement wheel – there is a relatively unknown variation comprising four wheels on the going side, with the extra wheel (equivalent to the centre wheel in an eight-day clock) driving the minute hand through the front-plate in the same arrangement as the centre wheel of an eight-day clock.

Such clocks tend to crop up at auction sales with no reference made to the fact they are only 1 per cent of all 30-hour clocks. The odd feature that often alerts a collector to the presence of the fourth wheel is a seconds hand in a subsidiary dial. It is extremely rare to see a seconds hand in an ordinary, three-wheel, 30-hour clock because the escape wheel rotates anticlockwise; so a seconds hand, if fitted, would likewise run backwards. There are collectors who specialize in four-wheel, 30-hour clocks, including birdcage examples from the late eighteenth century. Apart from their inclusion in *English 30 Hour Clocks* by Jeff Darken and John Hooper, they are not well-written up, which may stem from their still being 30-hour movements and the perception that they were (and still are) seen as a cheaply made, poor relation to the eight-day clock.

Actually, there is very little difference in the complexity, and hence cost, of a four-wheel, 30-hour clock against an eight-day type, so any argument about cheapness is difficult to justify.

EVOLUTIONARY PARALLELS IN EUROPE AND AMERICA

Dutch Longcases

Longcase clocks are not unique to the British Isles, there were well-established links between what is now the Netherlands and East Anglia and hence, London. It is natural that communications in the seventeenth century related to trade, and Dutch involvement in the land

A four-wheel, 30-hour clock. The hour wheel has been removed to show how the centre arbor is driven off the main wheel, in the same way as an eight-day movement.

20 *A Short Introduction to Longcase Clocks*

A Swedish longcase clock of the early nineteenth century. The movement is about half the size of an English 30-hour movement, with significant differences, especially in the method of count-wheel locking.

drainage of East Anglia led to wider cultural exchange. That connection was further reinforced when William of Orange, of the Dutch House of Nassau, invaded England in 1688 and subsequently eclipsed his father-in-law James II as king. Consequently, early Dutch clocks can be strikingly similar to their English counterparts and arguably are part of the early development of the English longcase.

Danish Bornholm Clocks – a Special Case

A British ship carrying longcase clocks ran aground on the island of Bornholm in the 1740s and initiated a longcase clockmaking tradition there, which persisted for many decades.

Other Scandinavian Clocks

The port of Tallinn is suggested as the centre of Baltic and Scandinavian clock and watch activity from the fourteenth or fifteenth centuries; but in Sweden, longcase clocks were produced at several locations, of which the town of Mora is often used as the name for such clocks.

Although the cases of Swedish clocks that were produced from the mid-eighteenth century are reminiscent in overall appearance to French Comptoise clocks, their movements are fundamentally different in layout. The typical plated, two-train movements are more reminiscent of an English 30-hour plated movement and it is likely that, through the Bornholm incident or some earlier evolutionary ancestor, they are related.

French Clocks

The rapid expansion of provincial clockmaking took place in the British Isles in the early eighteenth century, and a few decades later a similar parallel provincial clockmaking activity commenced in the Franche Compté region of France, near the Swiss border. That activity seems to have had parallels in the Black Forest area of Germany and, in each case, clockmaking developed into a cottage industry followed by farmers during the winter months. It is not surprising that the birdcage-type of movements

A typical French Morbier clock movement. Although the frame is reminiscent of an English birdcage clock, the wheel trains are arranged side-by-side.

characteristic of Franche Compté are quite dissimilar to English birdcage movements in wheel layout, but whether they are an evolutionary development, or just share a common ancestor, is not clear.

Although they were made as hanging wall-clocks, they were also housed in what is now, the instantly recognizable pot-bellied Comptoise or Morbier clock-case.

American Tallcase Clocks

The first longcases to appear in America were imports from the British Isles and, naturally, it is likely that American-made clocks were produced by immigrants. One way or another, an indigenous clock industry started and later developed into the mass-production of weight and, later, spring-driven clocks.

UNDERSTANDING CLOCK MOVEMENTS

To the novice, any clock movement seems to be an incomprehensible jigsaw of strange parts that can be assembled into just one solution, like a mechanical Rubik's Cube. Clock movements are ingeniously thought out and the layout of parts is the product of many years of gradual evolution from early clocks.

A good understanding of the way a longcase, or any, clock movement operates is one of the basic requirements for the repairer – more important than tools. In essence, the movement of the clock does two things: first, through the escapement, it maintains the uniform rate of oscillation of the pendulum by giving it a little nudge every swing; and, second, it mechanically displays the number of swings.

UNDERSTANDING ESCAPEMENTS

The swinging pendulum counts off the seconds consistently because it can only swing at the speed that is governed by its length. In order to maintain the motion, the escapement gives it a slight nudge at each swing, compensating for the natural mechanical losses.

Anchor Recoil

Almost all longcase clocks use an anchor-recoil escapement, which can continue to function despite quite heavy wear and misalignment. Eventually, if that wear continues unchecked, the efficiency of the escapement will inevitably fall to a point where it can no longer maintain the swing of the pendulum.

A healthy escapement will give the pendulum a noticeable supplementary swing, which can be seen by watching the motion and listening to the tick. The pendulum will be seen to continue swinging farther towards its extremity after the tick and, conversely, if the tick occurs just at the point of greatest amplitude, the clock is on the point of stalling and is probably in need of corrective work.

There are many variations in the detail but the general design persisted unchanged from its introduction in the late 1650s. It is particularly suitable for longcase clocks for several reasons:

- The small angle of displacement makes for simple regulation and good time-keeping because as a 'simple' (small angle) pendulum, it closely follows a basic law of nature: that in constant gravity, the time for one swing is only proportional to the square root of the length.

The entry pallet is to the left and the exit pallet to the right. As the pendulum swings to the left, the nib of the entry pallet rises to the point of the wheel tooth. The wheel is now at the point of escaping.

A Short Introduction to Longcase Clocks 23

As the pendulum swings farther to the left, the pallet lifts clear, allowing the wheel to turn clockwise until a tooth hits the exit pallet with an audible tick. The wheel has now escaped.

The supplementary swing forces the exit pallet downwards, which pushes the escape wheel backwards – that is the recoil. The pressure of the tooth on the pallet tends to check the supplementary swing and the pendulum reverses direction.

As the pendulum swings towards the right, the entry pallet falls and the exit pallet lifts to where the nib of the pallet is level with the tooth point. Again, the wheel is on the point of escaping.

As the exit pallet lifts clear, the wheel escapes until it is stopped by an entry tooth striking the entry pallet with an audible tick.

- The small amplitude of pendulum swing (typically between 4 and 6 degrees total swing, which equates to total pendulum displacements of between 75 and 100mm) has a direct bearing on the design and aesthetics of clock-cases.
- It is quite straightforward to make, set up and repair.
- It requires a fixed or stationary clock-case.

The term 'recoil' is often overlooked but the recoil effect is a necessary part of the functioning of this type of escapement.

Recoil is directly related to the supplementary swing that is necessary for the escapement pallet to lift clear of the escape wheel tooth. The effect of recoil can readily be seen in the behaviour of a seconds hand, which, having moved forward, regresses slightly after each tick. During the recoil phase, the escape wheel exerts a braking effect on the pendulum that counters the supplementary swing.

In most cases, the wheel teeth are cut far deeper than necessary, which makes the system immune to extreme swings of the pendulum. In other words, at the most extreme swing, when the pendulum bob contacts the inside of the case trunk, the back face of the pallet should be close to touching the back face of the wheel tooth.

Most early escape wheels have a hollow formed in the leading face of the teeth. That curve is not vital and in fact there is no one radius of curvature that must be used in designing the wheel.

Deadbeat Escapement

The year 1715 is usually given for the invention in London of the deadbeat escapement by Thomas Tompion's son-in-law, George Graham. The statement by Cesinski and Webster in their book, *English Domestic Clocks*, that 'It is incomparably superior to the recoil type, as far

The entry pallet is to the left and the exit pallet to the right. The point of the wheel tooth is in contact with the entry pallet impulse face. As the escape wheel rotates clockwise, the pallet is forced upwards, nudging the pendulum to the left.

After the wheel tooth has escaped from the entry pallet, an exit tooth strikes the check-face of the exit pallet.

As the pendulum reaches its extremity of swing and starts to return, the check-face of the exit pallet slides against the radial leading face of the wheel tooth, temporarily preventing any movement of the wheel.

The pendulum continues to swing to the right until the point of the tooth comes into contact with the impulse face, forcing the pendulum towards the right.

When the wheel has rotated sufficiently, the point of the tooth will align with the nib of the exit pallet. The wheel is now on the point of escaping.

as precision is concerned', is left without explanation, although one of the features of deadbeat escapements is the necessity to keep power on the going train during winding. (With a recoil-type escapement, there is a tendency for the swinging pendulum to drive the train in reverse direction during winding.) Consequently, deadbeat escapements are used in conjunction with some form of maintaining-power device in key-wound clocks. (In an endless rope, single-weight system, a deadbeat escapement will work satisfactorily without the necessity of a maintaining-power device because winding only relieves the pressure on the strike train.)

Semi-Deadbeat

The semi-deadbeat escapement looks like a true deadbeat but is actually a more precisely engineered variant of the anchor recoil designed to give minimal recoil. They are often found in later clocks that have the outward appearance of regulators, without having a maintaining-power device, and especially those clocks with long, centrally mounted seconds hands in which the effect of recoil would be magnified.

BASIC LAYOUTS OF THE MAIN TYPES OF LONGCASE CLOCK MOVEMENT

The Eight-Day Movement

The parts of an eight-day clock movement can be divided roughly into four groups:

1. A set of wheels and pinions comprising the going train mounted on spindles (clockmakers call them 'arbors') between the plates.
2. A parallel set of wheel and pinions that makes up the strike train.
3. An arrangement of wheels mounted on the front-plate that gears the hour hand to the minute hand, and also drives ancillary features like a date wheel or a moon-phase disc.
4. Finally, the strike-work that detects the appropriate moment to strike the bell and counts off the correct number of strikes for the hour.

The Eight-Day Going Train

The going side of the clock movement performs two quite separate tasks. First, it maintains the swing of the pendulum by giving it a little nudge at each swing – just enough to compensate for the losses caused by mechanical imperfections and air resistance. Second, it counts the seconds minutes and hours to 'tell' the time. There are just five moving parts in the going train of an eight-day movement.

The rotational force applied by the driving weight to the main wheel is transferred up through the train of wheels by a succession of gearing reductions from large brass wheel to small steel pinion. The main wheel drives the centre wheel pinion usually with a reduction

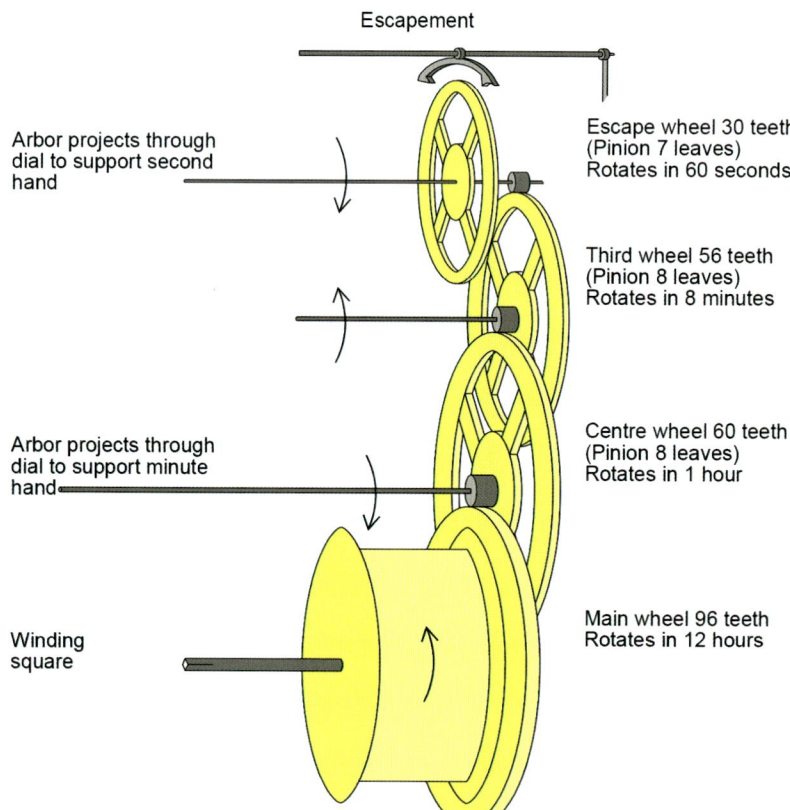

Schematic layout of the going train of a typical eight-day movement. (Steel: grey; brass: yellow.)

Schematic layout of the strike train of a typical eight-day movement. (Steel: grey; brass: yellow.)

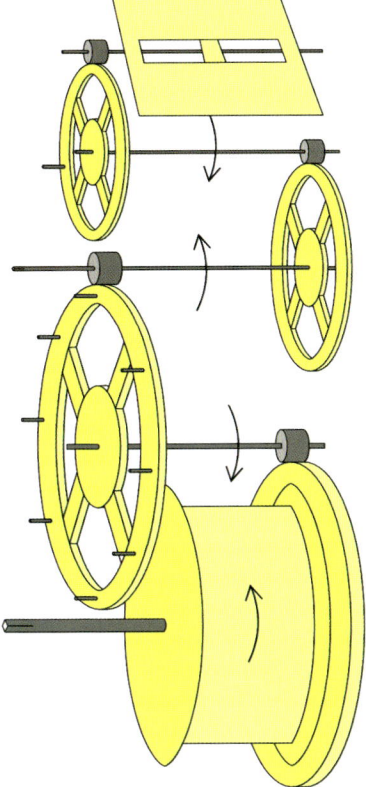

Fly – a simple air brake governor
Pinion 7 leaves

Warning wheel 48 teeth
Pinion 7 leaves
Single pin holds the strike cycle on warning

Pallet arbor projects through the front movement plate to support the gathering pallet

Pallet wheel 56 teeth
Pinion 8 leaves
Rotates once for each strike

Pins on the pin wheel trip the hammer
Pin wheel 64 teeth
Pinion 8 leaves
(same number of pins as the leaves on the pallet pinion)

Great wheel 84 teeth
Rotates once in 12 hours
(78 strikes)

of 12:1 so that while the great wheel rotates once in twelve hours, the centre wheel, which is connected to the minute hand, rotates once in one hour. The drive continues from the centre wheel through the third wheel to the escape wheel, which, as its name implies, allows the driving force to 'escape' at the rate dictated by the constant swing of the pendulum.

So the escapement rocks in time with the pendulum because it is directly connected. The two teeth or pallets of the escapement alternately stop the rotation of the escape wheel and simultaneously receive an impulse that is transferred to the pendulum. The distance between the pallets is usually 7½ teeth of the escape wheel. The exact number of teeth is not important but the ½ is: the 'tick' of the clock happens as the escape wheel, having been released by one pallet, is stopped by the one opposite.

The escape wheel moves by half a tooth at each tick, so a thirty-tooth escape wheel rotates in exactly 60sec and so provides the direct drive for a seconds hand.

The ratios of wheel teeth to pinion leaves, the tooth count, are not absolutely fixed; there are various combinations that give the same end result. Odd ratios and unusual layouts are more often found in early clocks.

The Eight-Day Strike Train

To clarify the terms, strike train means the arbors, pinions and wheels that make the hammer strike the bell. Strike-work means all the other devices that start and stop the cycle of bell-strikes.

The strike train also consists of five arbors; the wheels and pinions are arranged so that the driving weight falls at the same rate as that on the going side. One revolution (twelve hours) of the great wheel is required to strike the bell a total of seventy-eight times.

As the pin wheel rotates, the pins cause the hammer to lift against its spring. As the pin-follower on the hammer falls off the pin, the hammer springs back towards the bell.

Moving up the wheel train, the third wheel supports the gathering pallet of the strike-work and invariably, the number of leaves in the pallet wheel pinion is the same as the number of pins on the pin wheel. That coincidence implies that the pallet wheel, and hence the gathering pallet, rotates once for every time the hammer is lifted by the pin wheel.

Further up the train, the warning wheel (the most often misunderstood part of the movement) prepares the strike cycle in a first phase just before the hour and then, on the hour, releases the strike cycle proper.

Finally, in order to control the speed of the strike operation, the warning wheel drives the pinion of the fly, which is a vane-type of airbrake, usually rotating at roughly 2,000rpm and designed to even out and govern the speed of the strike-work.

An ordinary eight-day clock strikes nearly 57,000 times in a year, so like the going side, the deceptively simple design, which evolved over many years, has proved to be reliable and resistant to wear.

The Eight-Day Motion-Work

The motion-work consists of three wheels that transfer the drive from the minute arbor through an intermediate wheel to the hour wheel. The cannon wheel (usually of about thirty-six teeth) fits over and is driven from the centre wheel arbor. A slightly dished spring (the bow spring) sitting on a shoulder in the centre wheel arbor rubs on the inner face of the cannon wheel to act as a slipping clutch, so that the clock hands may be moved without moving the going train. The intermediate wheel (the reverse minute wheel) usually has the identical tooth count to the cannon wheel and drives the (usually seventy-two teeth) hour wheel through a (usually six-leaf) brass pinion.

The motion-work achieves a 12:1 reduction from the minute hand to the hour hand

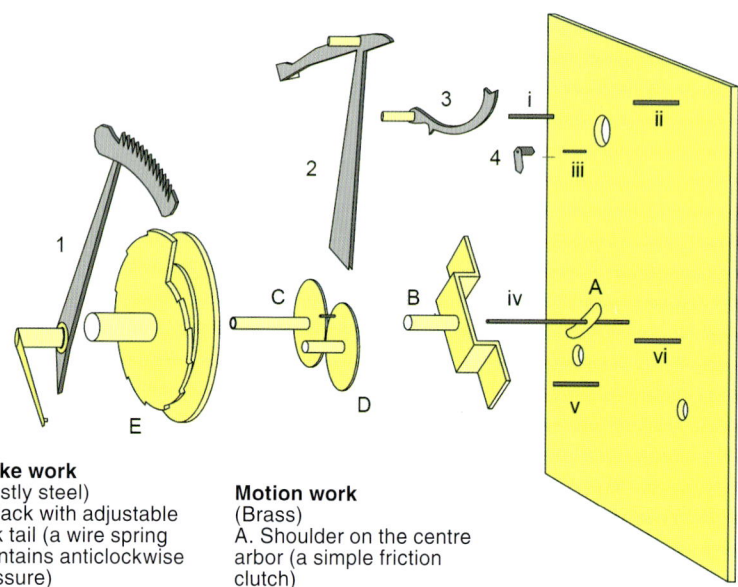

Strike work
(Mostly steel)
1. Rack with adjustable rack tail (a wire spring maintains anticlockwise pressure)
2. Warning piece with lifting tail that follows the reverse minute wheel pin
3. Rack hook; a ratchet that holds the rack after each turn of the gathering pallet
4. Gathering pallet; counts off the strikes by moving the rack by one notch per revolution

Motion work
(Brass)
A. Shoulder on the centre arbor (a simple friction clutch)
B. Bridge supports minute and hour wheels
C. Minute wheel
D. Reverse minute wheel with its pinion that drives the hour wheel with 12:1 reduction
E. Hour wheel with the snail, which controls the amount of throw of the rack

Posts and arbors
i. Rack hook post
ii. Warning piece post
iii. Pallet arbor a pin to the left stops the gathering pallet
iv. Centre arbor
v. Rack post
vi. Reverse minute wheel post

Schematic layout of the motion-work and strike-work of a rack-strike, eight-day movement. (Steel: grey; brass: yellow.)

but, in addition, a pin on the reverse minute wheel is used to activate the strike cycle. As the pin on that wheel comes to the top position, it lifts the warning piece to commence the strike cycle. The strike train starts to turn but is then stopped as the warning wheel pin stops against the warning vane. As the warning piece drops off the reverse minute wheel pin, the warning piece vane drops to allow the strike train to continue to sound the bell.

Date wheels and moon-phase discs are usually moved by separate wheels driven off the hour wheel.

Strike-Work

The striking of a bell at regular intervals was originally devised to indicate prayer times in religious houses and the Almanus manuscript, written in about 1480 by a German monk in Rome, contains descriptions of a number of weight and spring-driven clocks, including count or locking wheels that count off the correct number of strikes of the hour bell. By that time, the use of irregular canonical hours had lapsed in favour of the 24h system that we now use.

The first longcase clocks used the count-wheel system, inherited from those medieval clocks through their evolutionary progeny, the English lantern clocks of the early seventeenth century. In that evolution, the system of synchronizing the strike with the hour improved with the innovation of what we now call the warning system.

The rack-striking system was introduced during the second-half of the seventeenth century. By the beginning of the Georgian period (1714), most London clockmakers had adopted that system for eight-day clocks, although 30-hour movements retained the count-wheel system until the end of the longcase period. Because rack-striking is specific to the position of the hour hand, it allows for repeating work, which was popular until the mid-eighteenth century.

So, as a general rule (but with exceptions), eight-day clocks use rack-striking, while 30-hour clocks use count-wheels, with the count-wheel normally located behind the back plate, as was the rule with the earliest eight-day clocks. Some makers continued to use count-wheels well into the eighteenth century, and when they are fitted to later eight-day clocks they are usually found on the great wheel arbor between the plates, rather than behind the back plate, as is the rule for 30-hour clocks.

Count-wheel striking is described with 30-hour movements below.

This rack-striking system allows the hour to be repeated. Here the warning piece is extended to the right with a fixing hole for a pull-cord to activate the repeat. (Rack and gathering pallet removed.)

Rack-Striking

As described above, the connection between the going train (and hence the hands) and the strike train is made when the pin on the reverse minute wheel lifts the warning piece.

The warning piece lifts the rack hook and the rack flies back on its spring until the pin on the tail comes into contact with the snail.

The strike cycle is then free to commence, except that the vane of the warning piece obstructs the pin on the warning wheel until the reverse minute wheel turns enough for the lifting piece to drop off that pin. Once that happens, the warning piece drops and its vane drops clear of the warning wheel pin.

A typical eight-day, rack-strike clock. As the hour approaches, the strike cycle is momentarily released until the bend on the end of the warning piece, which protrudes through the opening in the movement plate, intercepts the pin on the warning wheel. As the lifting piece drops off the minute wheel pin, the warning piece drops and the strike cycle continues.

As the strike cycle continues, the gathering pallet draws the rack back to its closed position, counting off the strikes. Finally, the gathering pallet comes up against a stop on the rack, preventing further movement of the strike train until the rack is released at the next hour.

THE 30-HOUR MOVEMENT

The name '30-hour' is taken to mean all longcase movements having a single weight on an endless chain or rope, which is pulled to wind the clock. It is a widespread but not necessarily accurate name, and 'single weight' or 'pull-up winding' might be more appropriate because the duration or running time between windings is sometimes more than 30h.

The 30-hour movement is often considered to be a poor relation of the eight-day. Economy of construction is usually cited as the reason for their widespread production, but actually the difference in complexity and the parts involved is quite trivial. Aside from the winding drums and great wheels, an ordinary 30-hour clock has just two moving parts less than an ordinary eight-day clock, although it does have only one driving weight.

The great wheel/drum arrangement of an eight-day clock is more complicated to make than the equivalent sprocket and great wheel of a 30-hour clock, and surviving records seem to demonstrate that clockmakers charged more for eight-day clocks.

There is one strange difference between the layouts of 30-hour and eight-day plated movements that is hard to explain: in an eight-day movement, the going train is always on the right-hand side looking from the front, but in a plated 30-hour, it is always on the left.

The 30-Hour Going Train

Four-wheel, 30-hour clocks are a small minority and the common 30-hour going train has three wheels, often with quite heavy teeth.

The 30-Hour Strike Train

Almost all 30-hour clocks use the count-wheel system. The strike cycle is initiated in the same way as an eight-day, when the pin on the minute wheel lifts the warning piece. The warning piece performs three tasks:

- It lifts the warning vane into the path of a pin on the warning wheel.
- It raises a tag on a second arbor, which in turn raises a detent or locking latch from a slot on the circumference of the count-wheel.
- It raises a second detent from a locking arrangement on the second (hoop) wheel.

CLICKS IN GENERAL

The mechanical dissimilarities between 30-hour and eight-day movements extend to

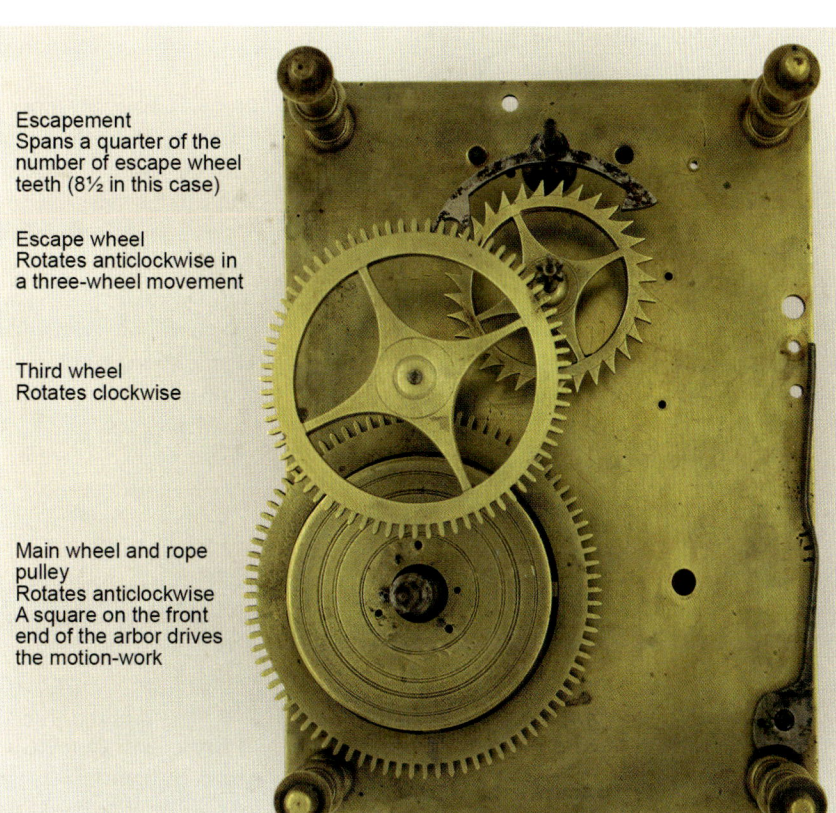

A typical 30-hour going train.

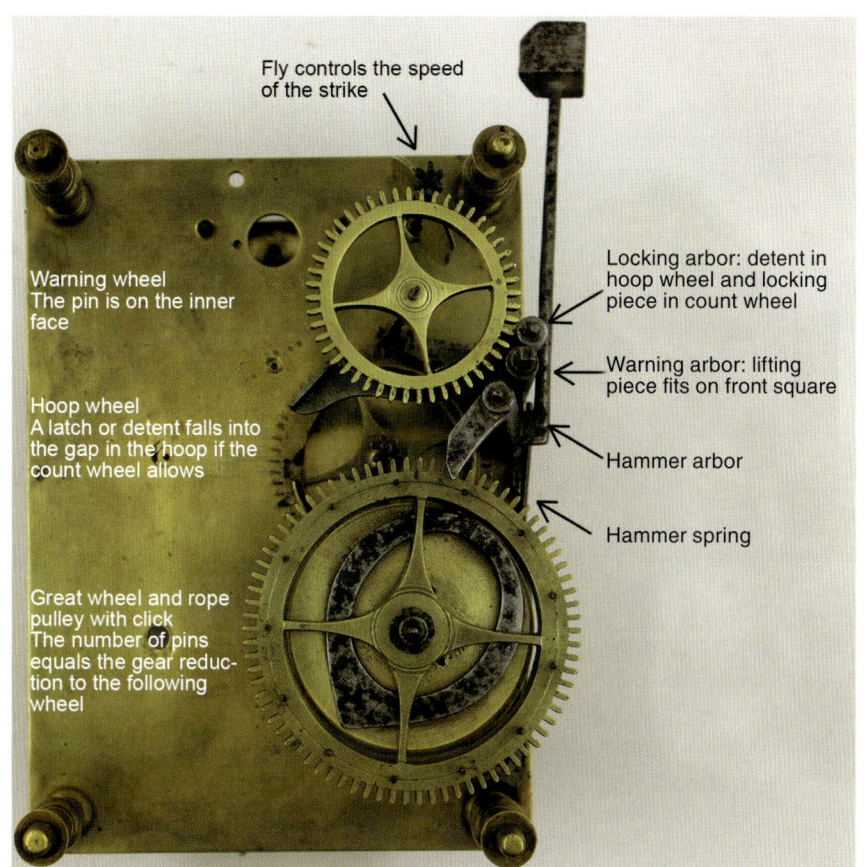

A typical 30-hour strike train.

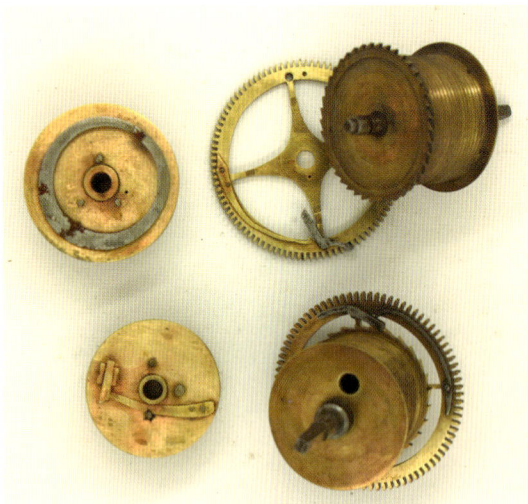

Eight-day clicks on the right: the upper one has apparently been repositioned. On the left, the two types of 30-hour clocks: earlier above and later type below.

the winding arrangements. The endless chain of a 30-hour clock means that not only a single ratchet is needed to raise the weight, but also, if the ratchet is fitted to the strike-side sprocket, then weight will always be applied to the going train.

Two distinct types of click are found in 30-hour clocks: the early type is a circular steel spring into which a step is cut, while the later type uses a flat strip spring and a pivoted pawl. In both cases they work on the crossings of the great wheel.

PENDULUMS

The pendulum is suspended from the back-cock by a thin, flexible strip of spring steel (the suspension or feather). The impulse that maintains the swing is derived from an extension of the escapement arbor (the crutch) and the rocking motion of the escapement is transferred to the pendulum by the sliding contact of a brass block of the pendulum within the jaws or forks of the crutch. It is essential that the slack between the crutch jaws and the pendulum block is not excessive, otherwise the pendulum cannot get the steady impulse it requires and the escapement will be vulnerable to damage or wear.

The back-cock, crutch and pendulum spring of an early inside count-wheel, eight-day clock. The location of the escapement pallet pivot below the back-cock implies the need for a sliding fit between the chops (the horizontal frame) of the crutch and the pendulum block. There should not be any excessive play.

The two weights on the left are lead 11lb (5kg); the cast-iron weights range from 9lb to 14lb (4–6.3kg).

The pendulum adjustment naturally adjusts or regulates the time-keeping of the clock, so a regulating nut is fitted to the bottom of the rod, which can be screwed upwards to speed the clock or down to slow it.

Most longcase clocks have steel pendulum rods, either steel rod or strip, but regulator clocks often use wooden pendulum rods with a slightly different crutch arrangement: instead of the usual open jaws that enclosed the pendulum block, the bottom end of the crutch has a horizontal rod that fits into a slot in the pendulum rod.

Lead weights were used in early clocks, often cased in brass sheet. By the late eighteenth century iron casting was more widespread and, as a general rule, white-painted, iron dial clocks were originally fitted with cast-iron weights.

Brass-cased weights continued to be used in good-quality clocks and also in later regulators and domestic regulators, not because brass enhances the time-keeping properties but, first, because regulators are made as objects of horological perfection and, second, because the trunk door of a domestic regulator is not intended to be opened during winding; the door is fitted with a glass window that renders the weights visible.

There is no formula for calculating the required driving weight because, apart from giving a slight impulse to the pendulum, most of the energy derived from the falling weight is used in overcoming the internal friction in the clock, the rubbing of meshing wheel teeth and of the pivots in their brass bushes, and that mechanical inefficiency is specific to any clock and its current condition.

More finely wrought, better quality clocks usually require less weight because the teeth of more accurately made wheel-work tend to roll together with less friction from rubbing. So, the finely made centre-wheel, single-weight clock made by Richard Stone of Thame runs well with a weight of about eight pounds, while twice that weight is often found in 30-hour clocks.

Replacing the original driving weight with a heavier one, or adding an extra lead collar, was a common form of 'repair' in the past and the presence of an excessively heavy weight may suggest an earlier substitution that is no longer necessary, with the wheel-work running as originally intended. An example of that is in the Samuel Buxton of Diss birdcage clock, acquired in an auction with its sixteen pound lead weight. Once the heavy wear in the bushes had been corrected, it ran perfectly with half that weight. A noticeably rapid repetition of the strike in a single-weight clock can often suggest the presence of a heavier than intended driving weight.

The driving weights of month-duration clocks (typically in the range twenty to twenty-eight pounds) tend to be roughly only twice as heavy as their eight-day counterparts, again suggesting finely made, low-friction wheel-work.

CASES

General Arrangement

The design details and appearance of the cases of longcase clocks follow an evolutionary path from the late seventeenth century to the beginnings of regional styles in the mid-eighteenth century, thereafter the production of longcase clocks was concentrated in the provinces and regional variations in style became more noticeable.

Although cases vary in their detail and the quality of workmanship, the general arrangement of construction is more or less uniform, and a few main structural elements are identifiable.

Backboard

The backboard is a vertical load-bearing element that extends the full height of the case, from floor level to the back of the hood, giving rigidity to the whole case and closing the dust-proof compartment that houses the clock. Mostly, the backboard is formed from one or two softwood boards, sometimes with flat transverse strips, but it is not unusual to find cases that use horizontal boards fixed between the sides of the trunk. Whether or not such horizontal boards are original or later replacements is usually a matter of guesswork, but sometimes there are clues such as wire instead of cut nails or circular saw marks on the wood. Horizontal boards are intrinsically unstable

Although cases were made from a wide variety of woods, softwoods were always popular. This country pine case dates from the 1760s when the Adam style swags were the height of fashion. It was only from the late nineteenth century that cases were made to look antique. Many pine cases painted to imitate exotic woods were unfortunately stripped in the mid- to late twentieth century.

because softwoods deform far more across the grain than along it, so there is little chance of arranging boards horizontally in such a way that they will transfer the load down to the floor.

Base

The base is made up of two architectural elements: a flat plinth, sometimes with feet, supporting the box-shaped socle.

It is essential that the base is resistant to any deformation, so it is made by fixing sheets of wood onto two rectangular frames, the top one of which, with a decorative moulding, is used to attach the base to the trunk.

After the backboard, the base is the part most likely to sustain damage, either through the effects of prolonged exposure to damp or from impact damage.

Base-Board

The base-board fits tightly into the base at plinth level to hold the base structure square and true, and also to prevent the ingress of dust. It is located in the part of the case that is most prone to damp and rot, and because it is never visible, its absence is often overlooked.

Trunk

The trunk is a three-sided structure, fixed to the backboard, that transfers the dead-load of the clock from the seat-board to the plinth and hence the floor. Structurally, an added complication is the necessary presence of the trunk door that gives access to the pendulum and weights. Consequently, the majority of the dead-load is supported by the sides and front corners of the trunk, so it is essential that the construction joints between the back edges of the trunk and back board are solid. The rigidity of the trunk derives from glue-blocks along the vertical joints, the top frame of the base and a second rectangular frame at the top of the trunk, which also supports the hood.

Seat-Board

The seat-board, as its name suggests, bridges across the upper extension of the sides of the

The hood of the clock pictured opposite. Crudely made but deceptively strong. The top consists of thin pine-sheets glued and nailed to the curved framework. There is no base-frame; the front lower plate is fixed directly to the sides.

trunk (the cheek-plates) to form a solid base for the clock movement. In later clocks, the seat-board is usually loose, just resting on, but occasionally screwed to, the supporting cheek-plates. Earlier clocks, especially 30-hour and, particularly, birdcage clocks, tend to have the seat-board fixed permanently. In the interests of preservation of historical evidence, a fixed seat-board should not be removed unnecessarily.

Seat-boards tend to be made from soft woods and can be quite prone to woodworm infestation and structural deformation. Obviously the sudden collapse of a seat-board would be quite a catastrophic event for the clock movement, so replacement may be a prudent step.

As a conservation or restoration exercise, a replacement seat-board should be made up from the same type of wood as the original and be a reasonable copy of the original without any attempt at artificial ageing.

Hood

Although the hood is not subjected to loads from the weight of the movement and driving weights, it is structurally awkward because it is open at the bottom and the back.

The structural base of the hood is a three-sided frame that rests on the top moulding of the trunk and is located in a sliding joint formed by the attachment of a wooden strip to each cheek-plate. An inner face behind the door fills the gap between the edge of the dial and the structure of the hood.

In a more conventionally made hood the side-plates are jointed to the three-sided base-frame. The return at the back of the side-plate fits tightly between the base-frame and pediment to give rigidity to the whole structure. Glue blocks are used extensively to reinforce thin joints.

Underside of the base-frame of an early type case. The tenons formed in the bottom of the side-plates are easily visible.

A later type case (1830s); all nailed and glued construction with the side-plates and the sliding joint fixed to the inner face of the base-frame.

The pediment, whatever its particular design, is attached to the side-sheets of the hood and the overall rigidity of the hood is provided by the attachment of the inner door facings to the side-sheets, and also by a return at the back of the side-plate.

The top of the hood is usually a flat sheet in square dial clocks but in arch-dial types it may be flat or barrel-shaped.

Early clocks, and especially chiming clocks, may have a pierced frieze incorporated into the pediment, the purpose being to improve the volume of the chimes. The piercings should be backed by silk, which prevents the ingress of dust, a little detail that is often overlooked.

Periodic removal of the hood adds to the wear and tear. It is surprising how often clock hoods are accidentally damaged after they have been removed to gain access to the movement.

Construction of the hood varies between mortised and plain glued joints. The standard method of fixing the side-sheets to the base-frame of the hood is to cut two, three or four mortises into the base-frame, but later cases, especially, are often of all glued and nailed construction.

Trunk Construction Details

All sorts of woodworkers made clock-cases, from country carpenters with very basic tools to the most skilled London cabinet-makers. It is usually possible to tell something about the case-maker from the complexity and construction detail of the case. The long-edge joints of the base and trunk are often simple butt joints in country clocks, but more complicated cases incorporate recessed or chamfered corners in the case-work. The stronger and neater mitre and stepped mitre joints used by cabinet-makers required skill and resources. Adding a chamfer or cant to each corner implies twice the amount of woodwork and suggests a level of sophistication in the woodwork more in keeping with provincial town cabinet-makers, rather than country carpenters.

The traditional and most frequent method of making the front of the trunk is to mortise

The edges of case-work can tell a lot about the skill and competence of the maker. The chamfer or cant in the base of this mahogany case and the quarter-round detail of the trunk edge suggest a level of expertise not usually seen in country clocks.

The most frequent type of trunk construction – the horizontal components that define the top and bottom of the door opening are jointed in to the vertical stiles.

This unusual reversed arrangement of trunk construction may be indicative of a maker working in isolation. It is rare in northern clocks.

the upper and lower cross-rails into the stiles, leaving an open frame for the door. There are exceptions and it is not just plain country clocks that diverge from the standard layout.

CHOOSING THE BEST LOCATION FOR A CLOCK

An often overlooked factor in the longevity of a clock is the choice of its location. An inappropriate location can have a serious impact on the life of a clock movement and/or its case.

Traditionally, the longcase clock was visible from somewhere inside the front door or otherwise in a 'good' room. The best place is in a safe, dust-free (away from open fires) position, where the air is not prone to strong changes in temperature or humidity. Consequently, the hallway inside the front door may not be the best location.

Once a suitable position has been found, the clock should always be fixed to the wall for stability and to prevent the possibility of it being nudged inadvertently.

One favourite but inappropriate place is the kitchen, where the atmosphere is likely to have variable humidity and airborne grease. Another used to be on a half-landing on the stairs; also not a good idea because of the likelihood of damage from passing luggage. It is surprising just how much damage is caused to a longcase clock by a fall down a staircase! But, wherever the clock stands, it should be on a solid floor, not carpet, and ideally, the backboard of the case should be screwed to a solid wall.

Mitred construction is rare and the joints often give trouble. This is found most frequently in southern or Midlands country clocks. In this case, the maker has formed a bead at the trunk edge to conceal the vertical joint and to enhance the overall appearance. He has selected a piece of well-marked quarter-sawn oak for the door.

Chapter 2
Dating Clocks

For the owner or collector, the ability to estimate when a longcase clock was made adds another dimension to mere possession, and collectors tend to take pride in their knowledge of the individual members of their collection. For the amateur repairer, whether or not he (or she) is also a collector, the ability to date clocks with reasonable accuracy is a part of the basic understanding that is so useful for satisfactory repair.

THERE ARE SEVERAL WAYS OF DATING LONGCASE CLOCKS, AND QUITE A FEW PITFALLS

Probably the most obvious way of estimating when a longcase clock was made is by researching the maker's name and place of work, if they are marked onto the dial. Makers marked their clocks for various reasons, legal and guild requirements and just as a way of advertising, but there are exceptions in all aspects of longcase clocks and anonymous clocks do turn up occasionally. There might be several reasons for the absence of a name: removal of an arch with its name boss from an early break-arch dial to convert it into a square dial (possibly reversing a previous 'modernization'); the maker may have been a journey-man or making clocks for others and, in some instances, it is likely that the maker's name and town were deliberately not applied when a clock was offered for sale in a rival town.

Names on clocks made in the British Isles, and especially longcase clocks, are quite well written up and there are various directories, books and internet sources of maker's details. Unfortunately, what ought to be quite an easy looking-up in a book exercise is often not so. Errors aside, a great many makers have yet to be included in the ever-growing lists and with given names tending to run in families, there is often confusion between different members

A clock dial typical of the early eighteenth century bears the name 'John Dent, London'. Researching the name yields several Dents but none who could have made this in the early 1720s in London. Much more research is needed because the lists of makers are far from complete. Plenty of dating evidence – the vestigial rings around the winding holes, the matted centre and the half- and quarter-hour ring inside the hour ring – are all early features, but the arch dial is later. The wheat-ear engraving around the name boss is a typical early London feature. The corner spandrels are Cescinsky and Webster type 21, common in London clocks from 1715.

Nigel's house clock: a third-period 14in dial inscribed 'John Fisher, Bilston', in a typical circa 1840 West Midlands oak case, which has been embellished with exotic veneers.

and cases that should be helpful in dating long-case clocks. The size, detailing and embellishment of dials, the styles of hands and, of course, the style and materials of cases, all provide useful evidence, but only if interpreted properly.

FIRST, SOME OF THE PITFALLS

In many instances, not all of the dating evidence in a clock is in agreement. There can be various explanations for inconsistencies: regional and local conservatism of styles, the individualistic approach of certain makers but, most significantly, after as much as three hundred years it is very unlikely that any clock can be found that is in its truly original condition. Some retain a good proportion of the original, while others definitely do not. It is only relatively recently that interested parties have accepted that the cultural heritage importance of early horological artefacts should, in principle, limit the amount of justifiable disturbance.

Movements crop up from time to time that are obviously major rebuilds of older clocks. It is equally possible to find early wheel-work in a late clock, and vice versa. Clock-repairers and sellers over the years have often adopted a pragmatic, profit-based approach to selling clocks, which is in conflict with the ideals of the purist connoisseur or collector.

Antique traders are often accused of passing off clocks that are not all that they are purported to be, but ultimately it is for the collector to examine all the evidence and draw his own conclusions.

The history of repair work that is evident in most clocks should be borne in mind when attempting to date a clock. Only a (perhaps quite small) proportion of alteration work is intended to deceive and it is quite likely that later work was done in a sincere attempt to extend the life of an old clock. So evidence of historical alteration work does not necessarily diminish the value of a clock, it may just add to the interesting history.

Cases are often replacements from lesser clocks, and a good many 30-hour movements were discarded when their cases were re-used for more valuable eight-day movements. Partial

or generations of a family, all producing clocks over many decades, often in different towns. (References to directories of makers and internet sources are given in the bibliography.)

The construction details of a clock movement can often give clues to its approximate date. Several reference books are available that give details of the changes in shape of the arbors (the clockmaker's word for spindles or axles) and collets, the brass sleeves that are soldered onto the arbors to support the wheels. In essence, later arbors are generally parallel and the collets are quite geometric or square, having evolved from the dome shapes as first innovated in the late seventeenth century.

Clocks are very frequently bought on the strength of their outward appearance but there are no two the same, and there is a clear progression or evolution of fashions in both dials

William Burton is recorded in Kendal from 1724 (d.1786). The spandrels are Cescinsky and Webster type 40, common from 1760 to 1785. The chapter ring is made for a single-hand clock (i.e. there are no minutes around the outside of the hours, only quarter-hours inside). But the clock has a motion-work and minute hand. The date aperture has been blanked off but looking at the front movement plate there is no trace of any date-work, so the likely conclusion is that the Burton dial was fitted to its present birdcage movement. (Actually birdcage movements are very rare in northern clocks of the second-half of the eighteenth century.)

or complete rebuilds are not unusual, where a restorer has taken a badly broken (what antique auctioneers call 'for restoration') case and returned it to use. Dials bearing the name of a good maker are often 'married' onto mediocre Victorian movements. Hands are quite often replaced with stamped modern reproductions.

Because clockmakers in London were governed and controlled by a guild, it has been suggested that the fashions displayed by London clocks of the early eighteenth century were rather blinkered. Consequently, dating London-made clock-cases, especially those made in the golden age between the late seventeenth century and roughly 1730, is quite straightforward, but similar market conditions and guild restrictions did not apply to provincial or country clocks, which often exhibit marked individuality.

One of the more troublesome alterations that later clocks undergo is a re-naming process, which is more common in white dial than brass dial clocks. Although the original maker's name and place of work are almost always marked onto a dial, in white dial clocks that marking is just black ink, which tends to rub off over the years. In the re-naming process, the original name and place are erased and replaced with something more commercially attractive, or otherwise a new name is added when the whole dial is repainted during restoration.

THE EVOLUTION OF LONGCASE CLOCK MOVEMENTS, DIALS AND HANDS FROM THE LATE SEVENTEENTH TO THE MID-NINETEENTH CENTURIES

In the early development of clocks in the seventeenth century, there seem to have been several almost simultaneous jumps: the introduction of the metre-long 'seconds' or 'royal' pendulum and the anchor escapement, combined with the innovation of both the plated method of construction and the eight-day movement; so that by the latter part of the century, both the 30-hour and eight-day versions of the longcase clock were well-established.

While the form and layout of seventeenth-century birdcage lantern clocks persisted beyond the end of the eighteenth century, it did so only as the 30-hour clock. There is a very clear evolutionary path from lantern clocks to early 30-hour clocks and, thereafter, only subtle changes in the construction details of the birdcage until their final disappearance in the early nineteenth century.

Although there are well-documented examples of plated frame 30-hour clocks from the late seventeenth century, birdcage eight-day movements, in the sense of arranging two trains of wheels, one behind the other and each with its own weight, are non-existent. There are possible explanations: first, a strong conservatism amongst clockmakers; second, the practical aspects of arranging two weights in that form to drive the two tandem wheel-trains would carry serious implications for the design of the case; and, third, key-winding is awkward in that configuration. So while plated movements seem to have been introduced in the 1670s as a means of achieving the side-by-side eight-day duration, the two distinct durations continued somewhat independently as alternative methods of construction. Indeed, from a practical point of view, for a single-weight clock, the configuration, whether birdcage or plated construction, has no implications for case design. There seems to have been only a gradual introduction of the plated frame for 30-hour movements into horologically isolated localities, especially in East Anglia and various other southern rural locations, which continued to use the older form.

Consequently, although birdcage movements tend to look antiquated, reminiscent of medieval clocks, it is not true to say that an evolutionary path exists from birdcage movements to plated frames. Plated movements represent an innovation jump in the late seventeenth century.

During the early beginnings, principally in London, longcase clockmaking was aimed at the wealthy classes. A gradual broadening of the scope of the market followed as the farming and industrial revolutions created new wealth, and domestic and international commerce, mining, canal building and manufacturing created a new class of consumers. A significant proportion of that new wealth was held in the British provinces, providing the impetus for a widespread proliferation of clockmakers throughout the cities, towns and villages of the British Isles.

By the 1730s, when longcase clock production was diminishing in London in favour of portable bracket clocks, the provincial makers, many of whom may have been apprenticed in London before returning to their native localities, enjoyed a huge expansion in business.

This dial, inscribed 'Thomas Grimwade, Fecit' (fecit is a Latin word meaning 'made it'), bears all the characteristics typical of London circa 1710 and it is reasonable to conclude that Thomas had served his apprenticeship in London before setting up in Beccles, Suffolk, where the clock was acquired. Several generations of the Grimwade family made clocks in Beccles and a William Grimwade is listed there as late as 1839.

The painting style of this clock by M. K. Barker of Beccles is hard to interpret because it includes both early and late features but, judging by the later (third-period) features, it was made some time after 1820. The absence of winding holes suggests a 30-hour movement, which, in general, cannot support a seconds hand, so this particular example is a late example of the minority class of four-wheel or centre-wheel 30-hour clocks.

Great wheels of two 30-hour clocks – the teeth of the circa 1700 clock (top) are longer and more slender (possibly through wear). The sprocket wheels also differ: the older example is decorated with turned rings, while the later one is plain and utilitarian.

Throughout the eighteenth century, production of 30-hour and eight-day clocks continued to provide for a wide range of budgets, but by the early nineteenth century production of 30-hour clocks dwindled, so it is unusual to find a late (post-1830) clock with a 30-hour movement. It is very unusual to find four-wheel (centre-wheel) 30-hour movements, alarm movements and other idiosyncratic clock types made during that late period.

Early provincial clocks mimicked the London fashions in cases and dials, but by the second-half of the eighteenth century, recognizable regional styles had developed from the distinguishable refined elegance of the early London styles.

THE EVOLUTION OF CLOCK MOVEMENTS

Evolution of Gear Tooth Shapes from the Medieval Period to the Industrial Age

It is known that by the mid-seventeenth century, the concept of a cycloidal curve was established within the scientific community, but it is not clear when the theory of meshing gear wheels developed and how clockmakers developed the skills necessary to design and cut the tooth shapes for gear wheels and pinions, so that the meshing teeth roll over each other without rubbing. Certainly, by the beginning of the nineteenth century the classical method of setting out the cycloidal gear form was documented. It is described in an 1802 engraving (plate IV) in Rees's *Clocks, Watches and Chronometers*. Throughout the nineteenth century the science of gear design developed further, with the widespread adoption of the involute tooth form towards the end of the century.

So the square-tooth profiles of pre-1650s iron clock wheels, which are reminiscent of those found in early wooden milling machinery, gave way to teeth with characteristically curved tops. It is tempting to interpret those approximately cycloidal tooth shapes in early longcase clocks as proof of an understanding of gear geometry. However, for the purpose of dating, long, thin teeth with pronounced square, rather than pointed tops, is evidence for an early clock.

Uniformity of tooth size through the wheel trains is another guide to great age, not a definite rule, but later clocks tend to use smaller tooth sizes higher in the train, with correspondingly smaller pinions and arbors.

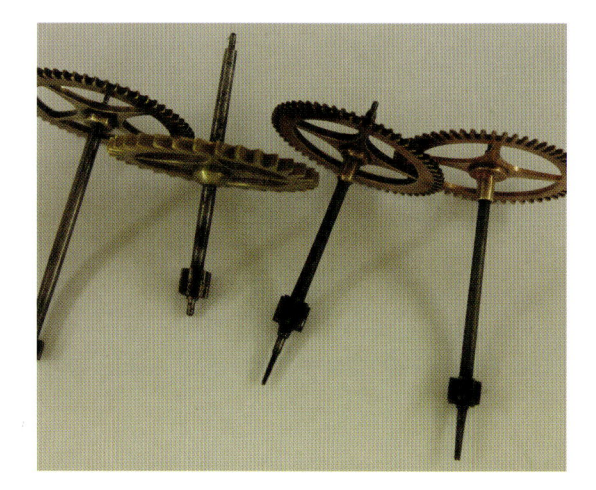

Evolution of wheel collets – older type on the left is silver soldered to the arbor with the wheel riveted in place. Second from left: a characteristic domed wheel collet from a circa 1700, 1¼sec clock. Third from the left: typical of the second-half of the eighteenth century. Far right: later, nineteenth century.

Other Dating Evidence in Clock Parts

Early clocks are recognizable by the shapes of their parts: the pillars supporting the plates, the arbors and the collets that hold the wheels onto the arbors. Early London clocks, and some early provincial clocks, have domed collets and their pillars tend to have characteristic fins turned into them.

The count-wheel of the striking work was originally mounted on the back-plate of both durations of movement; it remained in that position in 30-hour clocks but in eight-day movements it was conveniently relocated to a position behind the great wheel of the strike train.

The rack-striking was introduced in the 1680s as a replacement for the count-wheel system and, by 1710, it was the norm for London clocks.

From the late eighteenth century there was an expansion in production of regulator-type

The evolution of movement pillars. The carefully embellished pillars of early, particularly London, clocks (bottom) gave way to plainer styles in provincial clocks of the mid- to late eighteenth century (middle). Nineteenth-century pillars are even plainer and ultimately geometric (top).

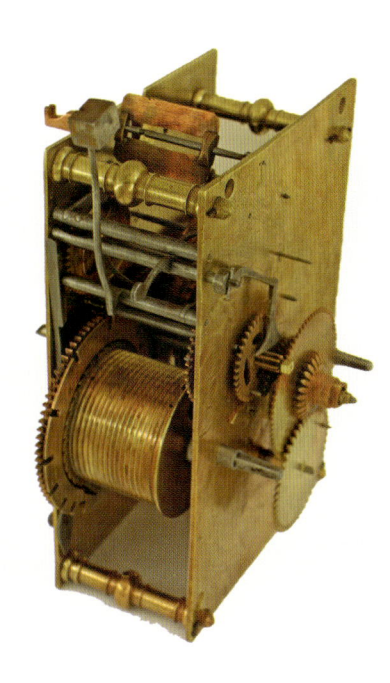

Inside count-wheel. Edward Barlow applied for a patent for his rack-strike invention in 1686 and from the early 1700s the count-wheel was almost exclusively used only in 30-hour clocks, but mounted outside the back movement plate. The inside count-wheel in this eight-day London clock is correct for the early style of pillars.

Rack-strikework – almost the rule with provincial eight-day clocks and always the exception in 30-hour clocks. A rack-strike does not provide useful dating evidence, except that the steel-work tends to be more curved and decorative in earlier clocks. In this mid-eighteenth century movement, decorative curves are incorporated into the rack hook and the warning piece.

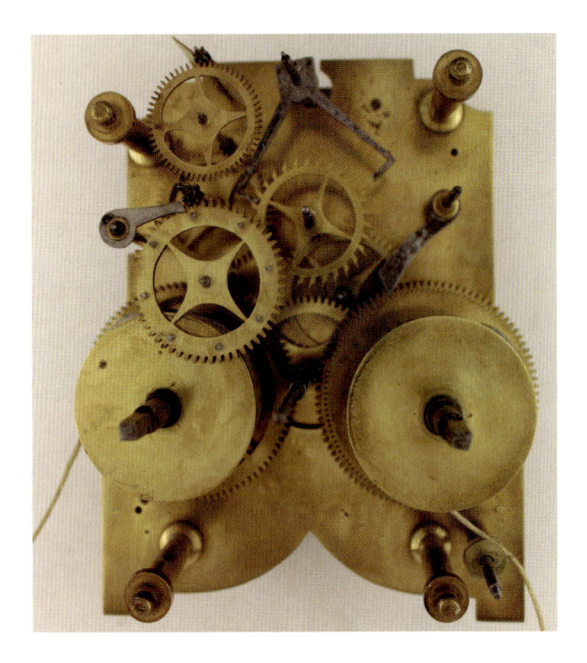

A Victorian domestic regulator, originally in White and Schopperle's clock and jewellery shop in Grafton Street, Dublin – plain pillars and geometric wheel collets date the movement to the mid-1800s. The shape of the escapement and the forward-pointing teeth of the escape wheel show this to be a deadbeat clock. Another feature of deadbeat clocks is the addition of a means of keeping power on the going train during winding. In this clock, the Harrison maintaining power is a toothed wheel and pawl behind the going side drum. (A piece of spring steel in the toothed wheel gives enough power to keep the clock moving for a minute or two when winding rotation of the drum removes the normal driving force.)

movements, which are generally larger with heavier plates, incorporating deadbeat escapements, maintaining-power devices and temperature-compensated pendulums. The 'domestic regulator', with its brass weights behind a glazed trunk door, was a common sight in high street watch and clock shops. They were often marked with the name of the shop.

The Evolution of Dials and Hands

After researching the maker's name and place of work as marked on a clock dial, the size and design of a clock dial itself is probably the next most useful means of dating, as long as there is reasonable confidence that the dial is original to the movement.

Brass dials with silvered chapter rings were the norm from the earliest longcases until the 1780s, and although the engraving details of the dial and chapter ring gradually changed over the years, the brass dial continued until a sudden style jump took place with the introduction of the white-painted iron dial. The changeover from brass to white-painted iron dials commenced in 1772 and culminated in the closing decade of the eighteenth century. (Naturally there are exceptions and throw-backs.) As a product of the industrial age, the success of the painted iron dial was largely due to the innovation of a universal adaptor, the false plate, which could be used to fit any clock movement to any dial (provided that the configuration of the winding arbors, the centre wheel and the seconds arbor were suitable).

It is a logical conclusion that from the introduction of the white dial, clockmakers laid out

A cast-iron false-plate made by Stansbie, Birmingham, fitted to the back of a second-period 12in dial. Stansbie is recorded as operating between circa 1805 and 1810, so it is reasonable to suppose that the dial and false-plate were supplied together. The dial feet are unalterable because of the painted dial but the false-plate feet would have been fitted by the clockmaker to suit his movement.

their wheel-work spacings according to the holes specified by the dial-makers, because once the dial is painted, it cannot easily be drilled. There is good evidence that painted dials were supplied, mostly (but not exclusively) from Birmingham manufacturers, complete with painted decoration in the corners and arch, and inscribed with the maker's name and town.

The presence or not of a false-plate is not a good indicator of age; they are not an absolute necessity, only a convenience for the clock-maker. Many white dial clocks were made without false-plates and they more or less disappeared after the mid-nineteenth century.

Early Brass Dials

The style progression of brass dials commenced with small, 8in-square dials in very early clocks through 10 and 11in-square in the late seventeenth century, to 12in-square by the early eighteenth century. The break-arch appeared in London clocks of the 1720s and was widespread by the decline in London longcase production after the 1730s.

A 30-hour clock by Thomas Henson of Yaxley, Suffolk. Although Henson is recorded in Yaxley between circa 1760 and circa 1780, the dial shows several early features: the matted centre, half- and quarter-hour marks and ringed holes that seem to mimic the winding holes and seconds dial of an earlier eight-day clock. The sad-mouth date aperture is a later feature in keeping with Henson's known period.

Spandrels were first formally described by Cescinsky and Webster in 1913; their original work has since been expanded and augmented and, although there are only about four dozen documented spandrel-designs, discoveries of previously unrecorded designs are always possible. Top left: C&W 8, popular before 1715 in London and later in country clocks. Top right: C&W 20 is widespread in country clocks up to the 1740s. Bottom left: C&W32, popular during the third-quarter of the eighteenth century. Bottom right: a slightly later rococo design, not described by Cescinsky and Webster, but recorded by Brian Loomes in 1985.

Provincial styles seem to have jumped back in time, so country clock dials of the 1730s are often similar in style to their late seventeenth-century London influences.

Plain, matted surface inside the silvered chapter ring evolved with engraving, which was first used to enhance the date aperture. The degree of engraving increased over the few decades from the 1720s to become more fanciful in the neo-classical tastes of the day. Rings around the winding holes of London clocks were introduced about the turn of the century and persisted for about twenty years but, again, the style is copied in later country clocks.

The designs of corner and arch spandrels were written up by Cescinsky and Webster in 1913 in their book, *English Domestic Clocks*, and that original work has been further enhanced by Brian Loomes and others. There are surprisingly few spandrel designs but they do follow a more or less ordered progression over roughly a hundred years, until the end of the brass dial period.

White-Painted Dials

White-painted iron dials were introduced in the 1770s, mostly but not exclusively made in Birmingham. For dating purposes, they were first categorized by Brian Loomes in his 1976 book *The White Dial Clock*. He set out three distinct phases of development of painting styles and the marking out of the hours and minutes.

- **The first period** – characterized by fine, delicately executed marking with Roman hours, Arabic minutes at 5min intervals and dots at each minute. The corner painting, replacing the earlier cast-brass spandrels, is usually finely executed with, typically, pink or red roses, blue cornflowers, strawberries and foliage. They are square or break-arch in shape and, where the arch is not used for any other purpose, it often contains a picture often referred to (inaccurately) as a vignette.
- **The second period** – more or less coincides with the introduction of startling new Regency fashions and is characterized by the replacement of Roman hours with Arabic. The corner spandrel painting became more abstract with Adam-style fans and there was a strong preference towards the break-arch shape. In general, the craftsmanship of the painting showed a marked deterioration when compared to first period dials.
- **The third period** – these dials are almost exclusively break-arch, often larger in size (14in width is not uncommon). The style of painting in the third period (commencing more or less with the accession to the throne of George IV in 1820) is characterized by a heavier approach; the minute numbers disappeared and the minute ring became a double line with radial marks for minutes, resembling a tram-line. The corner painting degenerated to landscape pictures of romantic ruins, often with a strong similarity in diametrically opposite pairs.

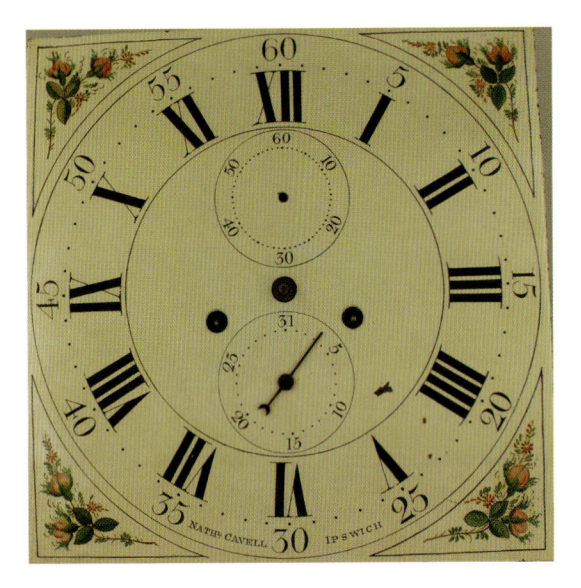

A typical first-period, 12in dial: very finely executed detail, often images of roses and strawberries with delicately drawn leaves. Dots mark the minutes with Arabic numerals at 5min intervals. First-period dials do not always have false-plates.

A typical second-period dial, 12in across with an arch. It retails much of the fine painted detail of the first period but with the minute numerals at 10 or 15min intervals. The big change is the adoption of Arabic numerals for hours. False-plates are the rule for second-period dials.

Third-period dials are usually bigger – up to 14in or more, usually with a break arch. The painting is heavier and cruder than in earlier dials. This early third-period dial shows vestiges of earlier designs and a typical romantic landscape in the arch. In later third-period dials, diagonally opposite corners are usually painted with similar romantic ruins. Minute marks became radial lines, often contained within two concentric circles.

Circular or Round Dials

Round dials gained favour at the end of the eighteenth century. They appeared as white-painted and also as engraved and silvered brass. It is not clear which was the first type but several researchers have suggested that under-employed engravers promoted the silver dials. For some reason, circular, white-painted dials are seldom decorated except for a maker's or retailer's name and place. They tend to have the austere functional look normally associated with regulators. Dating evidence is in the style of hour and minute marks, but they are usually of the third-period type. Silvered round dials seem to have died out quite soon but white-painted iron ones continued to be popular right up to the end of longcase production, in the mid-nineteenth century.

A plain, silvered dial, this one bearing the name 'Mullinex, Derby'. The single sheet of brass is engraved to mimic a composite brass dial. Neo-classical motifs are common in the central zone of silvered dials, in contrast to (mostly) strikingly plain, white-painted dials. (Mullenex is apparently a miss-spelling of Molyneux.)

Hands

Cescinsky and Webster recorded styles of clock hands (and a great deal more) in their *English Domestic Clocks*. However, as dating evidence, hands are unreliable because they are quite delicate and subject to damage, so they are often replaced.

In addition to functional but unsympathetic replacement of hands, clocks were occasionally 'modernized'; for example, with the addition of motion-work and a minute hand to early one-handed clocks.

In broad terms, the progression of longcase hands is from the single (hour) hand designs to non-matching pairs of steel hands, often intricately pierced and shaped before polishing and blueing; followed in the late eighteenth century by matching steel hands, again, pierced, delicately shaped and blued. In the early nineteenth century, steel was replaced by matching brass hands, which are quite often extraordinarily decorative. Regulator-type clocks are an exception; they are usually fitted with matched but

This 12in dial inscribed 'Thomas Benbow, Northwood', started life as the single-hand dial of a 30-hour clock. The rococo spandrels date it to the second-half of the eighteenth century (circa 1770) but it has been adapted to an eight-day movement by cutting winding holes (through the date disc) and attaching a clumsy pair of modern pressed steel hands.

The dial and hand of a classic 30-hour East Anglian single-hand country clock. Made in Diss, Norfolk in the late 1730s by Benjamin Shuckforth. There are no minutes – the inner divisions on the chapter ring are quarter-hours. The fine execution of the dial suggests a London apprenticeship.

plain hands, seldom pierced or decorated, the balanced seconds hand often mounted at the dial centre.

The Characteristic Shape of Early Single Hands

The very early London clocks of the mid to late seventeenth century were quite often fitted with only an hour hand. Without the benefit of a minute hand to adjust the time, the single hand was made with a return projection, so that it could be twisted more easily between the fingers with less likelihood of bending or damage. There is a clear evolutionary progression from the single hands in seventeenth-century lantern clocks to those in early longcases. The same styles were adopted in the country clocks.

Steel Hands

Throughout the brass-dial period, minute and hour hands were generally unmatched. The typically spade-shaped pierced hour hand differs significantly in style from the slender minute hand that was embellished by an S-shaped feature near the hub.

The fashion for matching steel hands coincided roughly with the introduction of the white-painted dial. The presence of original steel hands would usually suggest a date prior to the end of the second period of white-painted dials; the presence of matched steel hands on

A pair of unmatched steel hands from the mid-eighteenth century.

Dating Clocks

Matched steel hands became fashionable in the last-quarter of the eighteenth century – roughly the time when painted iron dials were introduced – so first-period clocks often have the earlier unmatched hands, while late brass dials might properly have matched hands. This Wigan dial by William Barker dates from the 1760s, so the hands are apparently a later replacement.

a third-period white dial could be taken as an historically incorrect later replacement. Finely pierced and shaped steel hands are difficult and time-consuming to make and polish.

Brass Hands

By comparison to steel hands, the brass equivalents of the early nineteenth century are relatively easy to cut and shape, require far less polishing and do not require to be blued. However, brass being weaker than steel, they are necessarily thicker and heavier. The heavy appearance is relieved by working details into the surface of the brass to create a more three-dimensional object that gives the impression of being light and delicate. The scope for ornamentation that brass hands offered the engraver led to the adoption of regional styles, so it is quite possible to find thistles worked into Scottish hands or shamrocks in the Irish equivalent.

THE EVOLUTION OF CASE FASHIONS

Broadly, the evolution of cases, and accordingly the dating evidence, runs from slender cases of walnut or oak supporting small, square dials of the late seventeenth century, to monumentally wide, heavy cases veneered in exotic woods supporting dials of up to 16in width. A typical third-period Midlands or Northern white dial clock has a 14in break-arch dial with the case 18in wide across the trunk and 24in wide across the pediment and base mouldings.

Brass hands are the rule on third-period dials. They are thicker and generally heavier because the brass is not as strong as steel. The often very intricate engraving breaks up the otherwise heavy appearance of a plain brass hand.

Regardless of lack of finesse in the execution, some country cases very clearly show the influence of earlier London designs. The elegantly proportioned Thomas Henson Suffolk case on the left compares favourably with the much older London Clock by Nathaniel and Thomas Chamberlaine.
(Steve Daniels)

Richard Stone of Thame is known to have served his apprenticeship in London and was working in Thame from circa 1770. The presence of Adam-style design motifs in the satinwood oval inlays of the trunk door and base suggests a date of the mid-1770s. The case is oak, rather than the more fashionable mahogany, and the hood with its barley-twist columns forming the sides of the door is more characteristic of the early eighteenth century.

The introduction of marquetry and Chinese or Japanese lacquer-work was largely, but not exclusively, confined to London-made clocks for more than one reason. First, the marquetry or lacquer work was expensive and accordingly against the trend of widespread, but less expensive, country clocks from the 1730s. Second, the brief period over which that style of case embellishment was fashionable mostly preceded the development of country clock production.

Several writers over the years have speculated on the reasons for the strange absence of mahogany in surviving clock-cases until about 1760. It is known that mahogany was first imported into England in the reign of George I, and by the 1740s was widely adopted in the furniture and cabinet-making trade. Clock-cases are the exception and the woods that continued to be used in London cases were primarily oak and walnut, the fruitwoods and occasionally exotics (often as veneers). Softwood was also used in casework and early examples were occasionally ebonized.

THE SHIFT FROM 'LONDON' TO PROVINCIAL AND COUNTRY STYLES OF CLOCKS – STYLE CHARACTERISTICS

It is natural enough that the very early provincial and country clocks closely followed the quite simple London styles of the day. It was within the abilities of country joiners and, at a small town or village level, the local coffin-maker, amongst others. However, as London styles became more embellished, a difference in design directions was established. By the end of

This large first- or second-period case stands over 98in (2.50m) high. The first-period 14in dial has apparently been repainted to commemorate the death of Nelson (1805). The dial retains the long trunk door and uses satinwood cross-banding against flame mahogany veneers of the front of the case.

the Queen Anne period (1713), London clock-cases had developed into marvellous exemplars of cabinet-making skills. The quality of the work involved in marquetry cases was hopelessly beyond the range of the vast majority of provincial joiners.

As the volume of the clock trade in the provinces grew in leaps and bounds in the first-half of the eighteenth century, so did case-making skills, so that during the second half of the century there was an increasing trend towards more embellishment. From about 1760, mahogany was used increasingly in provincial and country clock-cases, often as a cross-banding around the edges of doors, trunks and hoods.

Robert Adam returned to England in 1760 with design ideas and motifs, about the time Thomas Chippendale was writing his *Gentleman's and Upholsterer's Director*, a significant book of furniture designs. Clearly, the fashions of the day influenced the design of clock-cases but the Richard Stone clock-case contains some additional dating evidence: the inlaid oval is satinwood, which only became fashionable in the mid-1770s.

The fact that the Richard Stone case was very fashionable when it was made is entirely in keeping with the extremely well-made four-wheel, single-weight movement (which, although it looks like a 30-hour, actually has a duration between windings of four days). But, of course, not all clock-cases were made to up-to-date fashions and it is apparent that subdivisions in the market existed between town and country, and between affluent and less well-off. In addition, by the latter half of the eighteenth century, regional styles had begun to develop, moving farther away from the original London styles.

Although there are no absolute rules, the final phase of the evolution of clock-cases is very much one of regional styles. Scottish clocks of

An inappropriate marriage: an early third-period dial (Henry Kistler, Penzance) sits awkwardly in an early eighteenth-century case. The layout of the hood is early – especially the 'low arch' (in early break-arch dials the geometric centre of the arch was below the top of the square). The hood door here has the columns fixed each side and the hood has side windows (both early features). The early long trunk door extends almost to the hood moulding and the door-top mimics the shape of the hood.

the early to middle nineteenth century have a look of their own, as do Yorkshire and Lancashire clocks. Southern clocks tended not to grow to the extravagant widths of northern counterparts and are generally less flamboyant.

A CLOCK MARRIAGE EXPLAINED: IS THE CASE ORIGINAL? IS THE CLOCK MOVEMENT CORRECT FOR THE CASE?

Two of the most frequent questions about clock-cases are: 'Did the clockmaker make the case?' and 'Is the case original to the movement and dial?'

The answer to the first question is that, with the possible exception of country clocks put into simple pine cases, it is very unlikely that the clockmaker made the case. The craft system just didn't allow skilled men of one discipline to operate in a different craft. Making clock-cases is a skill in its own right, and the training and experience necessary to achieve a passable result is probably similar to that required to make the rest of the clock.

The second question is far more difficult to answer. Having established that practically all cases are made by others, it naturally follows that at some point the clock was 'married to a case' and some collectors seek reassurances that their clocks are all happy in their first marriages.

A third-period dial and movement housed in a modern case. Although the case is reasonably well-constructed and finished, it is a little subdued for the third period and might be more compatible with an earlier dial. It is largely a matter of personal taste with a new case and quite acceptable, as long as a replacement case has not been passed off as the original.

Today, the term 'marriage' has come to mean a much later putting together of a clock and a case. In the early 1900s, Herbert Cesinsky pondered on the possibility of clocks by good makers being put into attractive cases that originally housed less prestigious names. That practice may imply a fraudulent intent to deceive a buyer or it may have been carried out innocently by a person with a spare case and a homeless movement.

With experience, it is possible to spot when the parts of a marriage are of different regions or ages; in other words when the marriage is unsympathetic. Otherwise, if the style of the case is in keeping with the clock, then the marriage might be a successful union of two compatible parts.

When they are cared for, clock-cases tend to last indefinitely but they are prone to damage, rot, woodworm attack and, worst of all, are occasionally discarded when slightly damaged, in the belief that they are beyond repair. Consequently, the new replacement case is a special kind of marriage. Making a clock-case is far from impossible for a reasonably competent woodworker and easier if the traditional methods of construction are followed.

Most experienced woodworkers find it easier to make a good copy of an existing case rather than build from scratch. It is quite easy to make up similar components to similar dimensions resulting in a case that bears a strong resemblance to the original pattern.

Making a new case without a pattern to work from requires research and careful design to accomplish an aesthetically pleasing case that is stylistically correct for the clock.

Chapter 3

Materials, Tools and Equipment

MATERIALS

Brass

The terminology of metal alloys is complex, confusing and mostly beyond the scope of this book, but briefly, the brass alloys are mixtures of copper and zinc in various proportions, often with small amounts of other constituents that modify its behaviour for a range of applications.

Common or CZ108 brass contains 37 per cent of zinc, which is the greatest amount of zinc that can form a simple alloy; higher proportions of zinc result in more complex duplex alloys. To improve the machinability for components such as clock wheels, a small proportion of lead is added to a 60/40 copper/zinc mix, to make a free-machining alloy (typically CZ120).

Tempering and Annealing Brass

Brass can adopt different tempers. The old British Standard BS2870 classified tempers in a scale starting at the softest and most malleable state, annealed, through quarter hard, half hard, hard, extra hard to spring hard. Transitions up the scale are achieved by cold-working (hammering or rolling) and down the scale by heat treatment.

Brass comes in various forms: sheet, round bar and tube. Fifteen thou shim plate (15/1000th inch) is useful for springs and brass parts of strike-work.

Repeated hammering gradually increases the hardness, so making brass springs starts with cutting the shape from a piece of soft annealed brass and then hammering lightly on a flat anvil. The hard spring quality develops quite quickly but as it does, the tendency to cracking also develops.

When brass is heated to a temperature of 500–600°C the internal structure recrystallizes; to soften or anneal, the required temperature is much lower. The brass is heated until not quite hot enough to glow dull red (between 200 and 250°C) and once that temperature is reached it is left to cool, although quenching in water does no harm.

Working Brass

Brass is usually sold as half hard or hard, and machining work such as wheel cutting is not helped by annealing because the brass is rendered too soft to machine cleanly and the new teeth tend to deform as they are cut; it is preferable to use a free-machining alloy. During cutting, filing and machining, brass behaves differently to steel. Tools should be extremely sharp and, for machining, cutter speeds are generally significantly higher. Traditionally, new files were kept for shaping brass and then, as they dulled slightly, they were used for steel.

Casting Brass

Brass is a fairly easy metal to re-melt and cast in a sand mould, so as brass scrap accumulates from clock repairs it can be saved for the

The pattern is pressed into a mixture of damp, fine sand with about 7 per cent clay. The green mould is dried thoroughly before pouring the hot metal.

occasional requirement to cast replacement spandrels or similar. The melting point is approximately 900°C, within the range of a modest gas burner, and the results of casting will be good, especially with some added zinc.

For flat castings, such as spandrels, the mould is made by first mixing about 7 per cent clay with damp, fine sand and compacting into a box. The pattern is then hammered into the damp sand with a flat, wood block, and when the impression is satisfactory, the mould is dried. It is important to make sure the sand mould is completely dried or it may explode when the hot metal is poured.

Occasionally spandrels are of lead, which melts at a relatively low temperature. A small forge is ideal.

Steel

The development of industrial steel production from small-scale beginnings dates from the mid-eighteenth century with the invention of the crucible process by Benjamin Huntsman (who happened to be a clockmaker). Prior to that, relatively pure wrought iron was alloyed with carbon to make steel that could be hardened to make saws, files and other tools.

Consequently, the steel in early longcase clocks may be quite variable in quality but, by

Different tempers for different jobs. From dead hard files to soft and bendy. The die plate (top) is hard enough to cut threads in steel screws but not so brittle that the cutting edges will break: woodworking and engraving tools must be hard to the point that they are prone to chipping: screwdrivers must be hard at the tip but flexible in the shank. Clock hands and screws are heated to a dark, iridescent blue. Clock arbors and pinions must be hard but the escapement crutch needs to be malleable for easy adjustment.

the second-half of the eighteenth century, high-carbon steel (also known as silver steel, tool steel or gauge plate) was widely adopted. It contains about 1.2 per cent carbon and can be annealed to a quite soft material for cold-working and machining. However, its usefulness comes from its property of adopting a range of hardness and toughness by heat treatment.

In general, hardness is traded off against toughness for a variety of purposes, e.g. screws, cutting tools, springs, arbors and pinions, clock hands.

Modern mild and medium carbon steels containing about 0.4 per cent and 0.8 per cent carbon, respectively, are products of nineteenth-century industrial developments in steel manufacture and have no role in the restoration and conservation of early hand-made clocks. They are immune to hardening and tempering by heat treatment.

Working Steel

Steel is best worked in its annealed state, so the workpiece is first heated to dull red and then allowed to cool slowly. In that relatively soft state it is easily cut and shaped, and in general it machines well at quite low speeds, although cutting tip speeds tend to depend on the nature of the work and the material of the cutting tool.

Heat Treatment

While the annealing process (i.e. heating to dull red followed by slow cooling) renders high-carbon steel soft and malleable, rapid cooling from red heat (usually quenching in water) renders it extremely hard. In that quenched state, which traditionally is termed 'dead hard', it is so hard that it cannot be cut with a file; but that hardness is achieved only at the expense of extreme brittleness. That combination of properties is found only in steel files, which, of course, will shatter with a hammer blow.

Prior to tempering, the quench-hardening steel should be tested with a file to ensure that it really is dead hard and cannot be marked. Re-heating the steel relieves the brittleness, and the degree to which it is re-heated dictates the final properties.

The forge doubles as a brazing hearth by adding a few fire-bricks to retain the heat of the fire and to raise the temperature. Air from a converted car heater fan is blown into the fire via holes in the base-plate of the forge.

A methylated spirit lamp (left) is useful because the heat is localized and gentle. The blueing tray (back) filled with brass swarf allows steel hands to be blued uniformly. Soldering iron and blow-pipe in the foreground give precise and controllable heat for melting solder.

By a lucky quirk of nature, heating steel to the correct temper is quite easy to achieve because of the way that the colour of a polished surface changes with temperature. So, starting at the quench-hardened state, the surface is first buffed with fine emery or corundum paper and then, if a bright finish is needed (e.g. for a screw-head or clock hand), it is polished with rouge or chrome polish to a mirror finish.

As steel is heated, the surface develops a pale yellow colour, which gradually darkens through red and brown to dark purple, then dark blue and finally a paler mid-blue/grey.

Applying the heat uniformly is far easier with a smaller flame in a simple brazing hearth, rather than a large flame in the open.

The range of colours of heated steel shows its temper for different uses

Tempering Colour	Approximate Temperature Celsius	Tool type	Clock part
Pale yellow	220	Cutting tools for brass, hammers	Escapement pallets
Dark yellow	240	Milling and lathe cutters for steel	
Pale orange	240	Dies, punches, bits, reamers	
Dark orange	250	Twist drills, knurls, large taps	
Red	260	Wood chisels, large taps and dies	
Brown	270	Small taps and dies, woodturning tools	Strike-work, racks, lifting pieces, etc.
Purple	280	Cold chisels, centre punches	Pinions and arbors
Dark purple	290	Screwdrivers, springs	Clock hands (steel). Screws, hammer springs
Blue	310	Scrapers, spokeshaves	Bell stands

WOODS AND VENEERS

Softwood for replacement backboards, glue blocks and case reinforcement, and other parts not normally visible, is best cut from fresh timber. Good-quality close-grained timber with minimal knots is commercially available and is usually sold as red deal to differentiate from the fast-growing, open-grained white deals.

For the exterior of clock-cases, the variety of woods used is as great as the variety used in antique furniture generally. Consequently, a clock-case restorer might hold a large stock of scrapped antique furniture, such as wardrobes,

Scrap furniture, tables, wardrobes, old pianos, etc., also offer a wide variety of woods for restoration work.

A very wide range of veneers is necessary. Older veneers recovered from scrap antique furniture tend to be thicker than modern stock, which often requires a backing sheet to make up the thickness.

and other carcase furniture, tables and table-tops, and pianos for veneers and their timber framing.

Veneers are available from specialist craft suppliers but may be salvaged for re-use from scrap furniture. The method for removing veneers from old furniture relies on softening the animal glues using a combination of heat and moisture.

Step 1 Remove the Surface Finish
Any shellac polish, varnish or wax on the surface of the veneer should be removed with a paint stripper. The wood is then cleaned by scrubbing lightly with white spirit and fine wire wool. A light smear of boiled linseed oil will give an indication of how the surface will appear after it has been polished.

Step 2 Remove the Oxidized/ Weathered Surface, if Necessary
One of the advantages of using old veneers for restoration work is that they are more likely to blend in with other veneers, but certain woods fade in sunlight or take on a greyish tint with prolonged exposure to damp and air. It is easier to rub away the degraded surface while the veneer is still attached to its ground. Rub lightly with fine sandpaper (not coarser than 320 grit

The veneer is covered with wet cloth and heated with an iron. As the glue softens, a long blade is inserted to separate the veneer from the ground.

size) and inspect the results by rubbing the surface with some boiled linseed oil. If the result is satisfactory, remove the linseed by rubbing with a rag containing a few drops of turpentine and then finish with very fine sandpaper.

A cabinet-maker's scraper is effective in the right hands but only when it is sharp and even, with no snags or nicks on the edge.

Step 3 Remove the Veneer from the Ground
Cover the veneer with wet towelling and press down with a hot iron, starting at an edge. After a few moments, the glue will be soft enough to insert a long, thin blade between the veneer and the ground. Continue gently, alternating between heating small areas with the iron and inserting the knife-blade, taking care not to puncture the veneer with the knife-point.

Step 4 Remove the Old Glue from the Veneer
Turn the sheet of removed veneer glue side up and scrub lightly with a fine bristle brush and plenty of very hot water, flushing the dissolved glue from the inner surface of the veneer. It is important to remove all traces of old glue because when the veneer is re-used, the new glue should penetrate to the wood-fibres.

The cutting edge of a scraper is the bur, which is burnished into the edge. In use, the scraper is bent slightly to keep the corners off the surface.

Step 5 Dry and Store for Use

The moist veneer is placed between sheets of absorbent paper and weighted so that it dries flat. Walnut burr is notoriously difficult to prepare for re-use because the absorbed moisture tends to make very prominent distortions, which become brittle when dry. One way to overcome the problem is to soak the veneer in wallpaper paste and then, while it is still wet, press it onto a backing of silk or similar fine material. Once properly dried, pieces of veneer can be stacked for storage, preferably kept flat under a weighted plate.

CLOCK TOOLS AND EQUIPMENT

The Amateur or DIY Clock Restorer

For the amateur clock restorer, the term 'workshop' could mean the kitchen table or the corner of a spare room, and the variety of tools required will depend on the complexity of the work undertaken.

The gradual acquisition of knowledge and experience usually goes hand in hand with the acquisition of more and more tools, which become necessary for increasingly more challenging repair projects.

There are several milestones in the development from novice to expert of which the more significant are:

- Dismantling and successful re-assembling of a clock movement.
- Re-bushing worn pivot holes.
- Making simple parts (gathering pallets, lifting pieces, repairing broken wheel teeth, etc.).
- Making wheels and pinions.
- Setting out and making a complete clock.

Consequently, the list of tools suggested below is most of what the amateur longcase clock-repairer might gather up. After a few years of practice and learning they will be sufficient to make a clock from pieces of round bar, flats and sheets of brass and steel.

It is also most of what the professional repairer would consider necessary for day-to-day repair work.

Measuring tools: micrometers (imperial and metric), callipers, steel rules and depthing tool.

The depthing tool is used to set up the best depth of meshing of two consecutive wheel/pinion sets. The distance between arbors is adjustable using the lower screw and, once the best depth of mesh is found (the deepest meshing that still allows the arbors to spin freely), that distance is transferred to the movement plate using the pointed scribe ends of the runners.

- Measuring, designing and setting out the work:

    – Camera for recording detail.
    – Steel ruler.
    – Vernier callipers.
    – Micrometer.
    – Drawing equipment; paper and pens, etc. but preferably a personal computer with a vector drawing software package.
    – Scribes, dividers, gravers.
    – Depthing tool.

- Dismantling, manipulating and assembling:

    – Work bench.
    – Adequate lighting.
    – Eye-glasses (loupes) of various magnifications.
    – Bench vice.
    – Large and small hand vices.
    – A range of different sized pin vices.
    – Finger plate.
    – Various pliers, including clamping pliers.
    – A range of screwdrivers.
    – Cross and ball peen hammers and a clockmaker's hammer.

- Cutting and shaping metal:

    – Large and small hacksaws.
    – Piercing saw.
    – Files including needle files.
    – Thread taps and dies (BA and Whitworth in preference to metric) and screw-plates.
    – Broaches and reamers.
    – Hand drill.
    – Twist drill bits.
    – Centre drills.

Materials, Tools and Equipment 65

ABOVE: From small hand vice (lower left) to engineer's bench vice. The finger plate (front middle) is useful for working on wheels and clock hands.

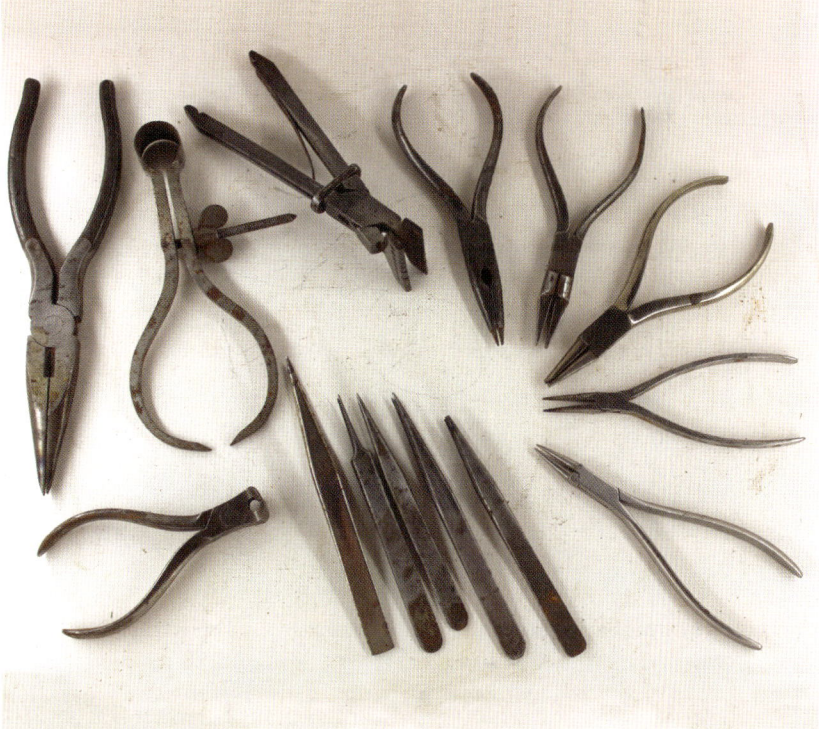

RIGHT: Tweezers and pliers of various sizes, side and end cutters.

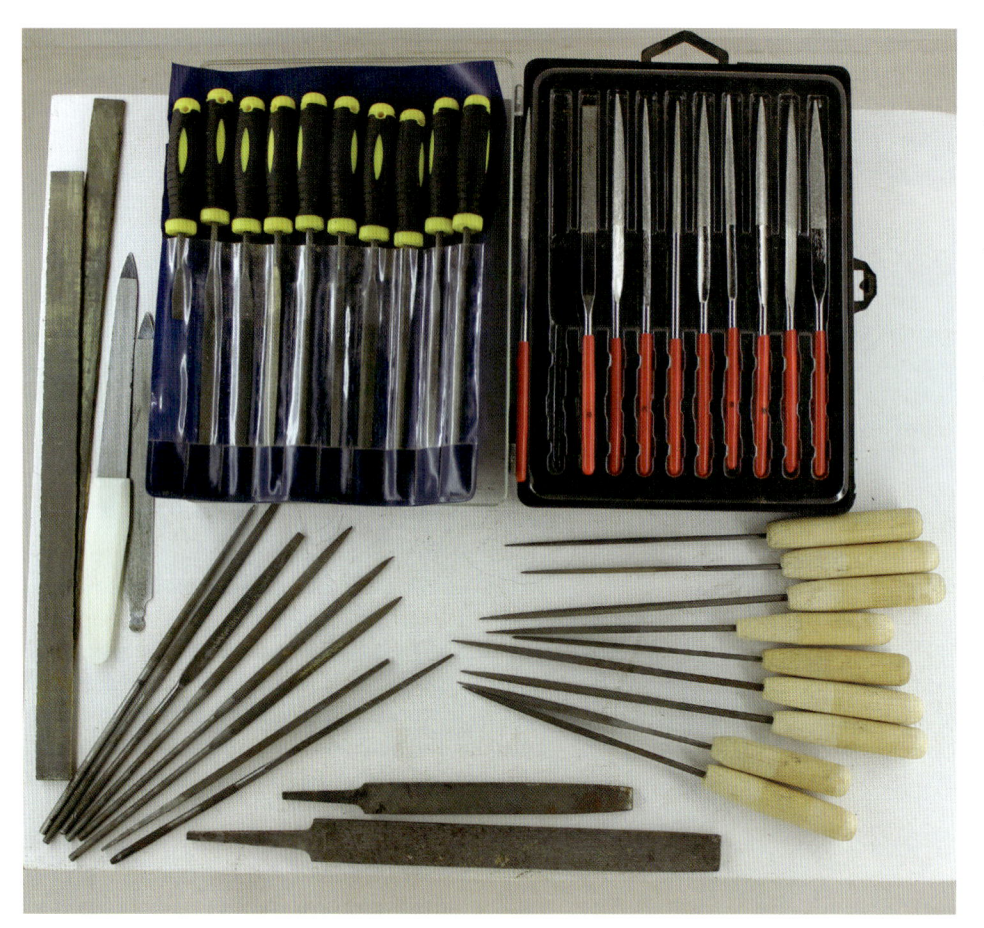

Steel files are obtainable from specialist suppliers. Diamond-coated files from craft suppliers tend to give a smoother finish. Emery buffs are made by gluing various grades of corundum abrasive paper to wood strips.

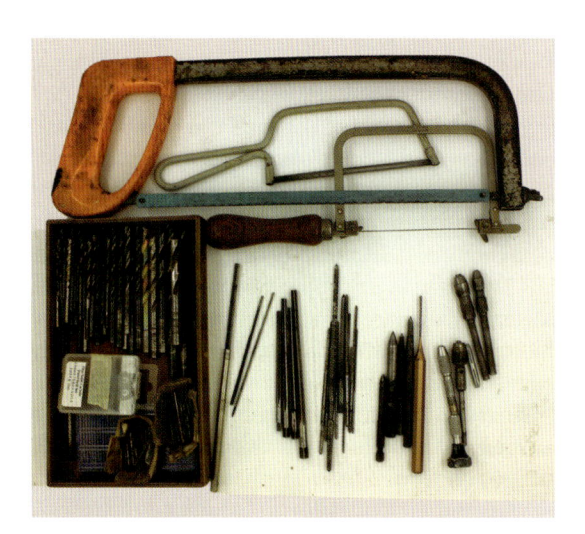

Hacksaw, junior hacksaw and piercing saw (top) and (from left to right) twist drills, burnisher, gravers, reamers, broaches, punches and pin-vices.

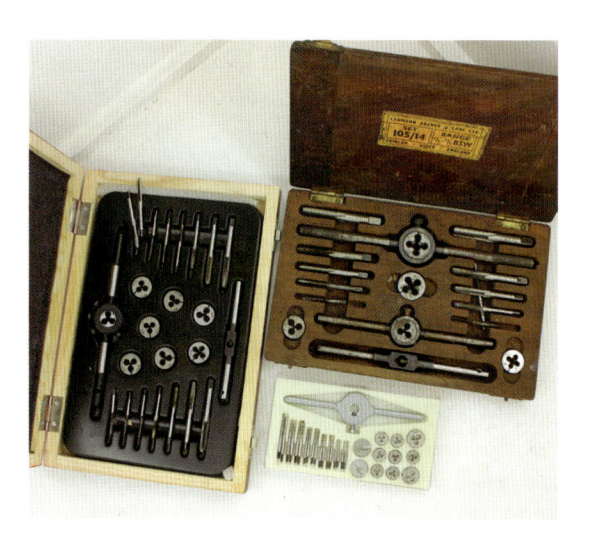

Thread tap and die sets; BA and BSW sets are useful for most later, longcase clock-work. For small threads, such as the fixing screw on later hour hands, a set of small metric taps and dies is useful.

Bench pillar drill (also know as drill press) and hand drills.

- All other metalwork:

  - Spirit lamp.
  - Small gas torch.
  - Brazing hearth (about a dozen fire-bricks and a gas burner).
  - Forge (for working iron).
  - Soldering irons.
  - Planishing hammer.
  - Engraving tools.
  - Bench drill press.
  - Clockmaker's or small engineer's lathe (preferably with vertical milling attachment), complete with cutting tools, dividing plates and so on.

A small engineer's or model-maker's lathe can be fitted with a three- or four-jaw or collet chuck. NB Remove the chuck key before starting the lathe. Very good advice is to use only a spring-loaded safety chuck key.

- Cleaning:
  - Hot and cold running water and sink.
  - Draining baskets.
  - Ultrasonic cleaning machine.
  - Hair drier.

- Miscellaneous repair tools:
  - Stakes and punches.
  - Jacot drum.

- Movement stand.
- Scribes.
- Variety of abrasive corundum stones and emery boards.
- One-off homemade tools for specific purposes.

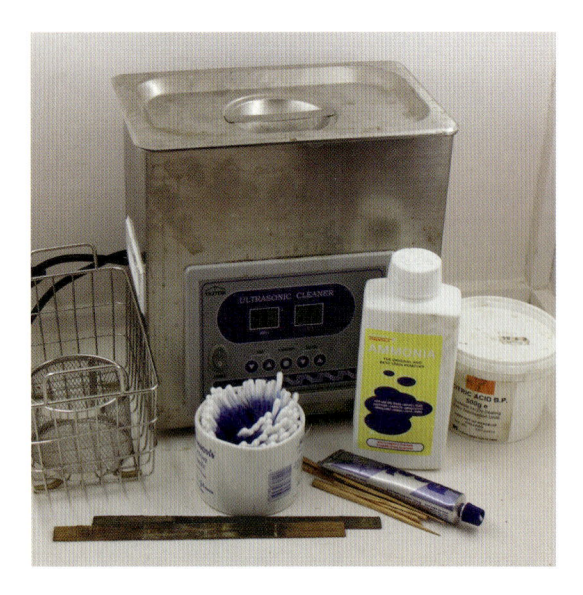

An ultrasonic heated cleaning machine is used with diluted citric acid and a detergent, or with unheated ammonia solution, but not both together. After ultrasonic cleaning, the superficial deposits are removed and the surfaces lightly polished. (As a conservation principle, polishing is undesirable because it removes original metal.)

Not forgetting boxes, cabinets and storage for all of the above to keep the work-place tidy, but above all, good lighting.

The Professional Clock-mender's Workshop

Whereas the amateur repairer may work on one or two projects at any one time, most professional repairers keep several jobs in hand at any time and, consequently, the workshop – apart from being more spacious – is divided into a number of discrete work-stations, each with its own work-bench and lighting.

Several individual work benches may be set out for different operations:

- Dismantling and re-assembly.
- Cleaning.
- Turning, drilling and milling.
- Bushing.
- Soldering.
- Test stands.

A homemade Jacot drum used to support a pivot while it is burnished. The clock arbor is set up in the lathe with the Jacot drum mounted in the tailstock chuck. The pivot rests in a groove on the drum, so that it is supported exactly on the axis of rotation.

The range of tools is roughly similar to those of the amateur, except that most professional clock-menders either inherit tools or spend a lifetime collecting them.

ABOUT JOINING METAL
Soft Soldering

After carefully cleaning and brightening both surfaces by filing or scraping, they are tinned by spreading soldering flux with a piece of copper wire and then heating until a little solder flows over the surface. Excess solder is wiped off with tissue paper before it solidifies.

More flux is added to the two mating surfaces and they are pressed together while heating, until the tinning flows. When cool, the work should be washed very thoroughly in hot water because the flux is acidic and, if not completely removed, it will initiate patches of corrosion.

Soft soldering is widely used in clockmaking and repair, and with care the heat applied will

A movement stand helps during re-assembly by positioning the movement at eye-level. This homemade stands used a car brake-disc as its base. The top revolves for easy access to the movement.

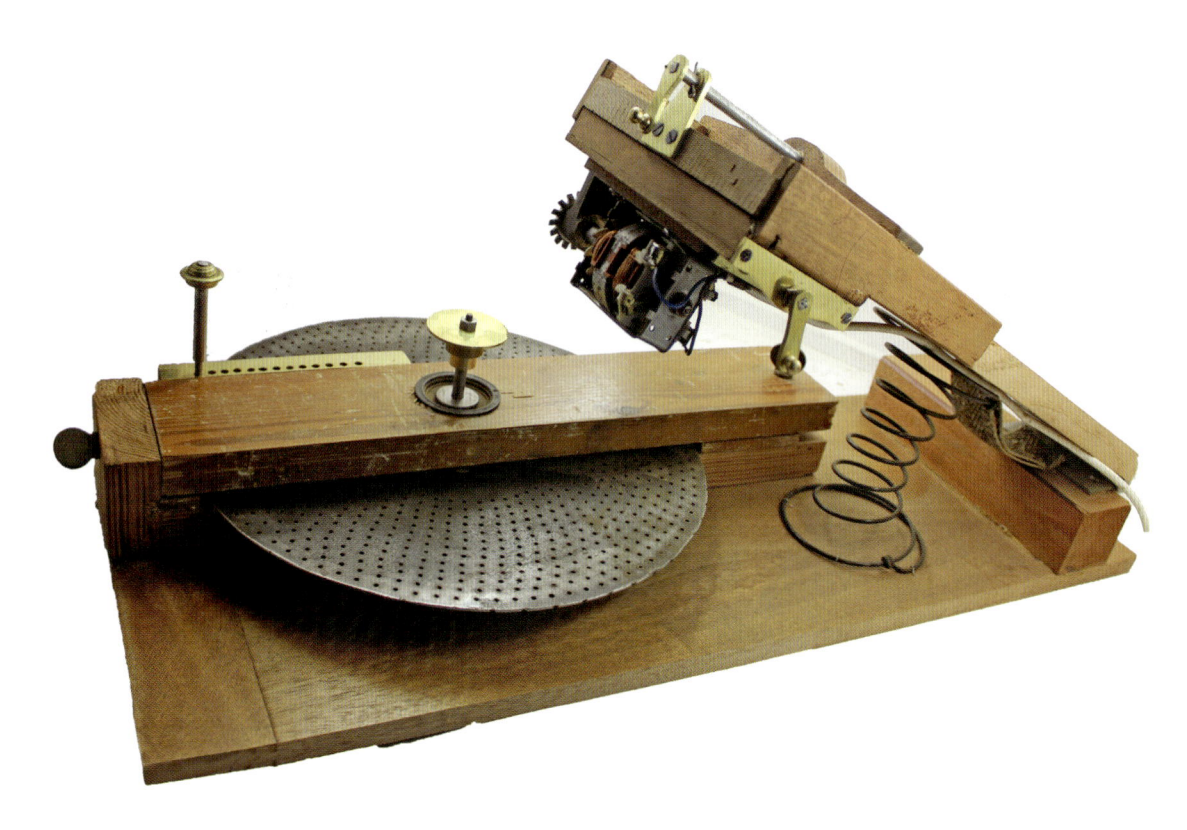

A homemade wheel-cutting engine with the motor cover removed. The wheel blank is mounted horizontally, fixed to the dividing plate arbor. A sewing machine foot switch operates the motor and a coil spring keeps the cutter clear of the blank. The cutting action is by pushing the hinged cutter deck downwards against the spring.

not exceed the blue-tempering temperature of steel.

Brazing and Silver Soldering

Soldering with melted brass or silver is a much stronger method of joining metals than soft solder but quite limited in its application to longcase clock repairs. The heat required exceeds the annealing temperature of both brass and steel.

To make a joint, both surfaces are filed clean and coated with borax paste flux. With a few grains of solder in the joint, the parts are bound tightly together and the whole heated with a gas torch. Both sides of the joint should be heated to the same temperature and once the solder runs, the heat is removed and after cooling in air, the joint is cleaned.

The silvering compound is rubbed onto the damped and salted surface; the black colour is colloidal silver.

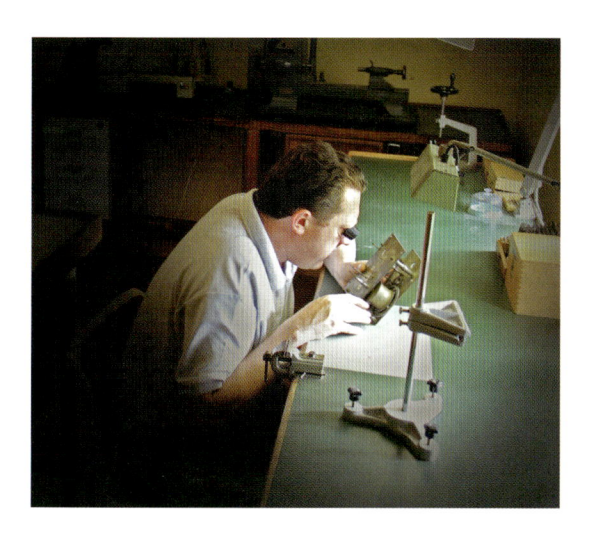

Good lighting is essential.

Silver solders, with a range of properties suitable for jewellers and clockmakers, are available from specialist suppliers and traditionally, scrap silver was used for steel joining steel.

FINISHES ON METAL SURFACES

Silvering (or French Silvering)

A cold, non-electrolyte process is used to apply silver to brass clock dials and especially their chapter rings. The resulting finish is microscopically thin and should not be polished; instead, it is protected by a thin coating of lacquer.

The brass to be silvered is cleaned and then rubbed with fine emery; chapter rings must be grained by using a revolving traveller. Then the silvering paste is applied with a soft, damp cloth. Silvering salts are available from suppliers or may be made up by dissolving silver nitrate in water and adding common salt. The precipitate of silver chloride settles to the bottom; it is washed and when it has again settled, the excess water is decanted. The white residue is mixed with salt and cream of tartar to make the silvering paste.

The brass is prepared by rubbing with damp salt and then the paste is rubbed all over in a circular action until a silver finish is achieved.

After thorough washing, the work is dried and lacquered immediately before oxidation can commence.

Lacquer

Lacquer is ordinary shellac polish, which is applied very thinly to polished brass or silver to prevent tarnishing. A quite dilute solution of shellac in alcohol is sufficient; if it is too thick, it will leave a streaked finish.

Solid brass parts are polished immediately prior to applying lacquer (French silvered parts should never be polished).

Austin's workshop. Plenty of flat surfaces and storage for tools and materials, with separate work-places for dismantling, making, cleaning, etc.

With care a spirit lamp is sufficient to blue a steel clock hand.

A gentle heat is applied and when the work is hand hot (about 60°C), the lacquer is painted on with a soft camel-hair brush and left to dry.

If the result is not satisfactory, the lacquer may be removed by dissolving in alcohol and repeating the process.

Warning Alcohol is flammable with a low flash-point – as with any other materials, read and understand the instructions and warnings.

Protecting Steel from Rust

There are several ways to protect steel from oxidation that leads to the formation of rust.

Polishing the surface is effective as a way of inhibiting rust, but once it has been pitted by corrosion, the affected area may require a protective coating with a traditional substance, such as juniper oil or shellac.

Wood-moulding planes, composite plane with blades and wood gouges. Gouges are ground either with the bevel on the outside of the curve (out-canel) or on the inside (in-canel).

Blueing by the application of heat is only suitable for certain steel parts because it alters the temper. It is ideal for clock hands and screw heads, but unsuitable for arbors and pivots.

WOODWORK TOOLS

Although cabinet-makers use a very wide range of tools for shaping and finishing wood, excellent results can be achieved by an amateur with a quite modest selection.

Clock-cases tend to contain various moulding profiles, which, when missing, can be reproduced with some basic tools, but a set of carving tools and some moulding planes are very useful.

Sharpening Edge Tools

The most important aspect of using tools to shape wood is the degree to which the cutting edges are sharpened. Feeling the quality of a cutting edge is best achieved by lightly sliding a finger or thumb across the edge (*do not rub along the edge*) – a jagged edge or bur will be easily detectable.

To sharpen a chisel, first put a few drops of light oil on the stone and rub the back of the chisel to remove any burs. Then turn the chisel over and rub the bevel. After a few strokes, it should be possible to feel a bur developing on the back of the edge. Turn the chisel over again and with the back flat to the stone, rub off the bur. Repeating a couple of times will not completely remove the bur – that is achieved by holding the chisel vertical and very lightly rubbing the bur away on the surface of the stone. Alternatively, a better result is achieved by honing or polishing away the bur. A metal polish, such as chrome polish or jeweller's rouge, is used on a flat, hardwood block, the back and bevel are rubbed alternately until the bur disappears and the bevel develops a mirror shine.

The back of the chisel is first rubbed back on an oiled stone.

With practice it is possible to get the angle just right.

The final stage of sharpening: the ground surfaces are honed on a hardwood block with a fine, chrome polish to remove the bur from the edge, leaving the blade surgically sharp and polished.

Chapter 4

Clock Maintenance

THE PRELIMINARIES – INSPECTIONS: WHAT TO DO AND WHAT TO LOOK FOR

In all mechanical devices, little problems tend to become big ones if not dealt with as they arise. As wear gradually shows itself, judgements are required about when intervention is necessary. Correcting problems as they arise will extend the period between major repair works and will probably mean less cost in the long term.

Apart from looking after the brass and steel of the movement, regular maintenance of a longcase clock also means looking after the woodwork of the case, with its hood and doors, hinges and locks. A badly neglected movement can be brought back 'from the dead', dials can, as a last resort, be repainted and replacement hands can be made; but a badly deteriorated case is apt to be (wrongly) discarded as useless or beyond repair. Although any case can be restored, complete case restoration is quite a serious undertaking and consequently expensive if put into the hands of a professional restorer. For the woodworker, whether amateur or professional, the task can be extremely challenging, especially on better quality cases that tend to be more intricate or ornate, and are often embellished with rare, exotic veneers. Consequently, if that sort of major rebuild can be averted by prudent early intervention, then not only there will be no serious impact on the value of the clock from modern case restoration, but the cost of that specialized work will be avoided.

REGULAR INSPECTION

Weekly

The regular re-winding of a clock is the ideal opportunity for a cursory check that all is well.

Winding 30-hour clocks naturally necessitates opening the trunk door in order to get access to the chain or rope to pull up the weight. It is also advisable to open the door of an eight-

While the trunk door of the 30-hour clock on the right must be opened to give access to the chain for winding, it is possible to wind the eight-day clock on the left without opening the trunk door – but not to be recommended.

day clock during winding because the pulley wheels should never be allowed to come into contact with the underside of the seat-board. During the winding process, there is an opportunity to follow the progress of the ascending weights, watching for irregularities in the gut-line and shake in the pulleys, and generally to listen to the action of the clicks (ratchets).

With 30-hour clocks it is a good idea to just keep an eye on the wear in the chain or rope, especially the spliced join in a rope or the individual chain links, which can stretch and pull open.

To prevent tampering with the time-keeping, some eight-day and longer duration clocks were made with a bolt that holds the hood door shut and access to that bolt is from inside the clock, just above the inside of the trunk door, which is normally kept locked.

Annual Inspection – Case Cleaning

If the atmosphere is reasonably clean and free from moisture, the clock-case should only need minimal maintenance, but the appearance of the polished wood of the case will deteriorate imperceptibly over the space of even a year, so cleaning and polishing the case with a good beeswax furniture polish is a very good idea. The annual 'spring clean' is an ideal time to polish the case and look closely for any snags or signs of change.

Furniture beetle infestation. The adults leave the wood in the summer months via fresh flight-holes and lay eggs in cracks and crevices. Early treatment is recommended.

The half-round beading around the trunk door of this early eighteenth-century clock is beginning to shrink and lift clear of the door. Without remedial action they will be prone to detachment and loss. (Steve Daniels)

Looking inside the case, as well as outside, the things to watch out for are:

- Changes in the structure – glued joints tend to come loose and wood tends to warp and deform. Shrinkage cracks and loose joints are symptoms of very low humidity, which is a feature of many modern houses.
- Loose, glued joints – old-fashioned wood glues that are made from animal hides and hooves tend to loosen in extremes of humidity (either very moist or very dry). Glued blocks that reinforce adjacent sheets of wood, especially on the inside of the trunk and the base, tend to come loose and fall to the bottom of the case. Although loose glue-blocks tend to get discarded, since the original maker had decided that they were necessary to help support the weight, it would be a good idea to reinstate them.
- Dry rot, wet rot and furniture beetle (*Anobium punctatum* or woodworm) cause irreversible damage to the wooden structure; it is possible to stop the progress of rot or woodworm and effective treatments are readily available.

A large gap between hood and backboard of this mid-eighteenth-century country clock. The gap had apparently been closed originally by wood strips fixed to the hood and, after they disappeared, the movement became inundated with dust and cobwebs, which eventually stopped the clock.

- Gaps in the case where dust could enter – it is not widely realized, but the case of a longcase clock performs several functions: apart from supporting the clock dial at a convenient height and being a significant piece of furniture in its own right, it importantly encloses the movement, protecting it from the harmful effects of dust. For the best protection from dust, the case should be as closely sealed as possible. The main problem areas for the ingress of dust are the top of the hood and the joints between the hood and the backboard of the case. The annual inspection provides a good opportunity to assess the build-up of dust on the top of the hood – and look at the places where it can get inside the case.
- Loose veneers, especially cross-bandings on the edges of doors and casework, are likely to detach completely if not fixed back into place. The loss of even a small piece of veneer can have a serious impact on the appearance of a clock-case and it is only with experience that the difficulty of exactly matching a veneer becomes apparent. In general, the more 'complicated' the pattern and figure in the veneer, the more difficult it is to match.
- Doors tend to deteriorate quite quickly if hinges and locks are not in good working order. If doors are difficult to open or have a tight spot, they are likely to develop more serious problems if not attended to. The trunk and hood doors are intended to be kept properly closed to minimize the ingress of dust into the case and hence into the movement.

DIALS AND HANDS

Dials are actually quite prone to damage from enthusiastic cleaning. With any clock dial, the best advice is usually – don't touch. Brass dials are mostly embellished with quite fragile cast brass corners (spandrels), which are held in place with small screws or rivets; they are very easily snagged with a cleaning cloth. Early

The loss of black ink from this third-period dial is quite typical of over-cleaning and frequent time-setting. The losses from the hour numerals are concentrated just above the radius of the hour hand, where a fingertip has been rubbing the dial while adjusting the minute hand. The maker's name and place have almost disappeared and can now be read only with difficulty in ultraviolet light. Careful re-touching could be the best option.

spandrels might still retain their original gilt but more often retain a fine covering of shellac lacquer, which prevents tarnishing; in either case they should not be rubbed.

Earlier brass dials (up to the introduction of painted dials) have a separate brass chapter ring onto which the hours and minutes are carefully hand-engraved. Originally, the chapter ring of a brass dial clock would almost always have been silvered, and that extremely thin layer of silver should never be polished because once the underlying brass is exposed, it will continue to tarnish. Likewise, late brass dials, from the late eighteenth century, are made from a single sheet of brass thinly silvered and engraved directly. Re-silvering a chapter ring or silvered dial to restore its original appearance is not a difficult procedure but it is made much more difficult where the chapter ring has been repeatedly polished to the extent that the original concentric graining is obliterated; it requires reinstating before the re-silvering can commence.

All clock hands are delicate and prone to damage, so the best advice for regular maintenance is – don't touch, until absolutely necessary. Early hands are cut from steel and blued by heating in brass swarf or filings. The blue effect is quite easily worn away by polishing but, in a damp atmosphere, it is also susceptible to corrosion with the development of rust patches. If rust has taken hold, it may be stopped by the application of thin oil such as juniper, but the corrosion cannot be reversed without re-polishing and re-blueing.

In the case of major damage to steel hands, making replacements is possible with much practice; otherwise they are available from specialist makers or, as a last resort, cheaper, stamped replacements are widely available.

Later hands (from the early nineteenth century) were made from brass and, although they are usually thicker than the earlier steel types, regular maintenance should be restricted to a light brush with a soft camel-hair brush. All hands are prone to damage, especially when they are being pushed around the dial to set the time. In the case of breakages, a soft solder mend might suffice but for serious damage, brass hands are probably easier to make than repair, and good results may be obtained with practice.

MOVEMENTS

There is no quick way to achieve regular maintenance on a longcase movement without a significant amount of dismantling. Moreover, it is a mistake to apply oil liberally when a clock becomes irregular – that sort of often well-intentioned but misguided maintenance is detrimental to the clock movement; the film of oil captures dust, which, apart from binding the works, will greatly increase the rate of wear. Aerosol lubricants are particularly unsuitable because they leave a sticky residue, but in any case, clock wheels should never be oiled. The only place for oil is on the pivots, where they are supported in the holes in the brass plates, and a very small smear on the escapement pallets. Otherwise, the movement should be kept dry and clean by sealing it as far as possible in its case.

Movement maintenance entails proper dismantling, inspection, cleaning and re-oiling every few years, as described below. The exact frequency of cleaning and re-oiling will depend largely on individual conditions but anywhere between three to ten years is about right.

The steel hands of this first-period clock are finely cut but the style uses 'thorns', which are very easy to snag.

REMOVING THE MOVEMENT AND DIAL FROM THE CASE

The hood is the part of the case that most protects the movement from dust-fall. It is removed to gain access to the movement and dial, but while they are still attached, the pendulum and weights prevent the movement from being lifted clear of the case. The pendulum is removed after first removing the driving weight(s).

Step 1 Removing the Hood

The hood almost always slides forward on runners: It would be extremely unusual, but not impossible, to find a clock-case where the hood lifts instead of sliding forward. Such clocks date from the early beginnings of longcase clocks but were often converted to forward sliding. Clock-cases with lifting hoods probably belong in museums.

Prior to attempting to slide the hood forward, it is advisable to check for the presence of a bolt on the inside of the trunk above the door opening.

The hood door has been opened temporarily during hood removal to show the sliding joint where the runner on the hood base moves relative to the guide on the outside of the case up-stands.

Having dropped the bolt (if present), the hood should be free to slide forward. The usual method is to grasp the sides of the hood and gently work it forwards. Sometimes a slight side-to-side rocking motion is helpful.

Once the hood has been removed, it should be put somewhere safe; that sounds obvious but a great many hoods have been damaged by leaving them in somebody's way.

Step 2 Remove the Weights

30-hour clocks have one weight and are more properly called single-weight clocks. Eight-day clocks have two weights, and musical and quarter-chiming clocks have three. Unhook the weight (warning – heavy!) from the pulley and again find a safe place to leave them. In addition to being easy to trip over, the cylindrical sort tends to roll. It is a very good idea to mark the weights with a piece of chalk, especially where there are three.

For eight-day clocks, it is usually easier to wait until the weights are at their lowest point (i.e. just when the clock is ready for re-winding). The reasons are that, first, it gives an opportunity to examine the gut-lines and, second, when the lines are wound onto the drums with no weight attached, they are apt to spring out and wrap around the arbors and drums in a haphazard way.

Take care when removing the weights – if the movement is not fixed securely, it may topple backwards under the weight of the pendulum.

Step 3 Remove the Pendulum

The pendulum hangs from the back-cock by a thin strip of spring steel, the suspension spring or feather. At the top end of the suspension spring, a brass block is located in a notch in the chops of the back of the back-cock. About two or three inches below the point of suspension, a brass block fits quite closely between the chops of the crutch.

Examine the crutch to establish whether it is an open or closed loop.

Put one hand through the trunk door and grasp the pendulum rod. With the other hand, gently lift the top of the suspension spring up and back to clear the back-cock.

If the crutch has an open loop, disengage the pendulum crutch block by sliding it backwards clear of the crutch. Otherwise, if it has a closed loop, lower the pendulum gently through the loop and withdraw it through the trunk door.

The crutch arrangement on regulators usually comprises a horizontal pin projecting from the lower end of the crutch, fitting into a slot in the pendulum rod. The procedure for removal is the same as for open-loop crutches.

Step 4 Check the Relationship of the Movement to the Seat-Board and the Case Cheek-Boards

Loose Seat-Boards Eight-day clock movements are usually fixed to a removable seat-board, which might be screwed or nailed to the cheek-boards of the case. Gentle, lifting pressure on each end of the seat-board will establish whether or not it is free. If it seems to be just resting on the cheek-boards but doesn't lift, check for nails or screws in the ends of the seat-board. As a general rule, remove fixings that are obviously a later addition but don't disturb those that look original. The style of the case may give some clues but early eight-day cases (made before the mid-eighteenth century) occasionally have fixed seat-boards.

Fixed Seat-Boards The most common type of clock to have a permanently fixed seat-board is the birdcage-type single-weight clock where, for earlier examples, it is the rule rather than the exception. As with any aspect of longcase clocks, there are many exceptions and eight-day clocks by country makers very often have fixed seat-boards. When a country maker was producing far more 30-hour than eight-day movements, it follows that the case-maker would follow the same pattern or style of making for each clock type.

Whether the seat-board is fixed or loose, plated movements are invariably fixed onto it by two bolts, which are either straight, screwed into threads cut into holes in the two bottom pillar, or with hooks that loop over the pillars.

With the pendulum and weights removed, a loose seat-board is lifted clear of the case up-stands, taking care that the line pulleys do not snag as the seat-board and movement are removed.

Although it has been disturbed recently, the seat-board of this unconventional eight-day clock was fixed securely to the case up-stands. Slots for the gut-line and the hooks on the holding down bolts are visible. The very heavy guide blocks fixed to the outsides of the up-stands are a recent addition.

The movement is set on two loose blocks on an assembly frame. The vertical position is easiest to check for general condition.

Step 5 Remove the Movement and Dial

If the seat-board appears to lift without obstruction, the dial, movement and seat-board assembly may be lifted clear of the case, taking care to bring the gut-lines and pulleys up through the top of the case. Otherwise, find the bolts that hold the movement onto the fixed seat-board and unscrew the fixings to allow the movement and dial to be lifted clear of the seat-board.

Once the movement and dial, with any gut-lines, rope or chain along with associated pulleys, have been completely removed, the whole assemblage should be placed on a couple of wood blocks on the work-bench to keep the movement upright, preventing it from falling backwards onto the crutch.

SEPARATING THE DIAL FROM THE MOVEMENT FOR INSPECTION OR REPAIR

The dial obscures the front of the movement and so prevents a thorough inspection. Once the hands have been removed, the dial may be detached from the movement. Clocks made during the period 1770s to 1830s tend to have a false-plate between the dial and movement. That plate, a sort of universal adaptor, was introduced with painted iron dials to allow any movement to be fitted to any dial. They seem to have disappeared towards the middle of the nineteenth century but the presence of a false-plate on a brass-dial clock may suggest a clock that is the result of a marriage of parts.

Step 1 Remove Hands

The minute hand is usually located on the square end of the minute wheel pipe. A tapered steel pin and a washer hold the minute hand onto the centre wheel arbor. There is a slight pressure on the washer from the friction bow-spring behind the minute wheel. Once the pin is withdrawn, the minute hand pipe will move forward slightly on the arbor as the pressure is taken off the spring. The minute wheel may then be lifted clear of the dial.

The hour hand is located onto the shoulder of the larger pipe of the hour wheel, which is concentric with the minute arbor; the hand is often held on with a small screw or a pin. For rack-striking clocks, the relationship of the hour hand to the hour-wheel (or, more accurately,

The hour hand of an eight-day clock is located onto the hour wheel pipe with a small screw. This hand has been adapted by cutting a notch for the screw 180 degrees from the original hole.

the snail) is crucial; in count-wheel striking clocks there is some flexibility. Hence, screws or pins are more likely to be encountered holding hour hands on eight-day than on 30-hour clocks. In older clocks a square in the hour hand may be a tight or interference fit on the hour wheel pipe.

Seconds hands are invariably carried on a pipe that fits over an extension of the escape wheel arbor, so removal is simply by pulling free with a slight twisting action.

Where date hands are fitted, they are attached to the date wheel mounted on the back of the dial, usually by a small screw. Consequently, it is unnecessary to remove the date hand.

Step 2 Removing the Dial

Plated Movements A dial made for a plated movement is provided with three or four post-like dial feet, which are pinned, either directly to the front-plate of the movement (brass dial clocks and, later white-painted dial clocks) or alternatively (in first and second period painted dial clocks) to a false-plate, which is itself pinned to the front-plate of the movement. The tapered steel pins are pulled with a pair of pliers or, if too tight, may be tapped loose with a long, flat-ended punch. With the pins removed, the dial may be lifted clear. Where a false-plate is present, the pin-removal procedure is repeated to allow the feet of the false-plate to be withdrawn from the front-plate of the movement.

Birdcage Movements A dial made for a birdcage movement is slightly different in the manner of fixing to the movement; instead of the stepped cylindrical dial feet common to plated movement dials, they are usually fitted with up to four right-angled flat lugs, which are pinned onto the movement. The pins are withdrawn using a pair of pliers and the dial is lifted clear of the movement. Once the dial has been removed, the movement is ready for a first inspection and, following that, if necessary, further dismantling to allow a more detailed inspection of the internal wheel-work to establish the need for cleaning and oiling or repair.

BACK OF THE MOVEMENT – REMOVING THE BACK-COCK AND BELL-STAND

The back-cock performs two roles: it supports the back end of the pallet arbor and also supports the top of the pendulum. Ideally, the bushing hole in the back-cock that supports the pivot at the back of the escapement pallet arbor will be level with the point of flexure of the suspension spring (i.e. very slightly below the chops).

Since the back-cock and escapement pallet arbor/crutch should be removed for close inspection, it is as well to do that before removing the peripheral parts (the motion-work and strike-work) from the front-plate. The advantage of that order of work is that once the back-cock and crutch have been removed, the movement may be laid down on its back, flat on the work-bench, without having to use large blocks or movement supports.

Tapered steel pins through the dial feet are the usual way of securing the dial to the movement. On early good-quality movements, latches were sometimes used instead of pins. The screw visible below the dial foot secures the chapter ring and the screw at bottom-right secures one of the corner spandrels.

The back pivot of the escapement arbor is supported in a bushing in the back-cock. The chops of the back-cock are directly over the oblong frame at the bottom of the crutch. Removing the bell-stand and back-cock will allow the movement to lie flat on the bench.

With the back-cock removed, free the front pivot of the escapement arbor, invert and lift to draw the crutch through the hole in the back movement plate.

If the bell-stand is screwed to the back face of the back plate, removal is convenient for later handling but not vital for inspection.

Plated Movements

Step 1 Remove the Back-Cock

The back-cock is fixed to the back plate by two screws (typically about 3BA) but there are usually two steady pins pressed through to accurately locate the back-cock. Remove the two fixing screws and pry the back-cock from the plate, taking care to keep it reasonably straight until it releases the back pivot of the pallet arbor.

Step 2 Remove the Escapement Pallet Arbor Assembly

With the back-cock removed, it is possible to manipulate the pallet arbor/crutch to remove it though the back plate. The usual procedure is to rotate the arbor by half a turn and then turn it upwards, while drawing the crutch forwards through the back-plate.

Birdcage Movements

Although, it is often easier to work on birdcage movements in the upright position, the pallet arbor must nevertheless be removed for close inspection, so it is as well to remove it, complete with the crutch, sooner rather than later. The fixing screws for the back-cock, which also supports the escape wheel arbor, are screwed vertically downwards to the top plate. They are removed to allow the back-cock and escapement pallet arbor/crutch assembly to be lifted clear of the movement.

FRONT OF THE MOVEMENT – REMOVING THE PERIPHERALS

The motion-work takes the drive from the centre-wheel arbor, which carries the minute hand, and through a 12:1 gearing arrangement, drives the hour hand. Because of its location and physical connection with the internal wheel-work, the paraphernalia of the motion-work and strike-work impedes thorough inspection of the parts of the movement between the plates – the two trains of wheels that are the heart of the clock.

The motion-work is held in place by a bridge piece screwed to the front-plate. The wheels and pinions of the motion-work rarely suffer from the sort of chronic wear that is liable to prevent smooth working. It is set up to trip the hourly strike, usually by the action of a lifting piece that rides on a pin on the reverse minute wheel.

The gradual wearing of rubbing surfaces in the strike-work is apt to show itself in a general slackening, which may result in slop or backlash in the hands and erratic chiming behaviour (such as occasionally chiming an extra strike or losing synchronicity).

Rack-Striking, Eight-Day Plated Movements

Step 1 Remove the Hour Wheel with Snail and the Rack

It is often necessary to remove the rack simultaneously with the hour wheel because of their close relationship. The rack spring should be disconnected from the projection at the lower end of the rack arm. The rack is usually held onto its post by a small tapered steel pin pushed into a transverse hole near the top of the post. Removing the pin will allow the rack to slide off its post, while sliding the hour wheel, complete with its snail, off the pipe of the hour bridge. The rack spring should be removed at this stage to prevent damage during handling.

Step 2 Remove Lifting/Warning Piece, Rack Hook and Gathering Pallet

There are two more parts of the strike-work mounted on posts above the centre of the front-plate. The lifting/warning piece and the rack hook are usually retained on their respective posts by transverse tapered steel pins. The gathering pallet is retained either by friction fit to the square extension of the gathering wheel arbor or, occasionally, by a small nut or even a small tapered pin.

Step 3 Remove Motion-Work

Two screws (usually about 2 or 3BA) hold the

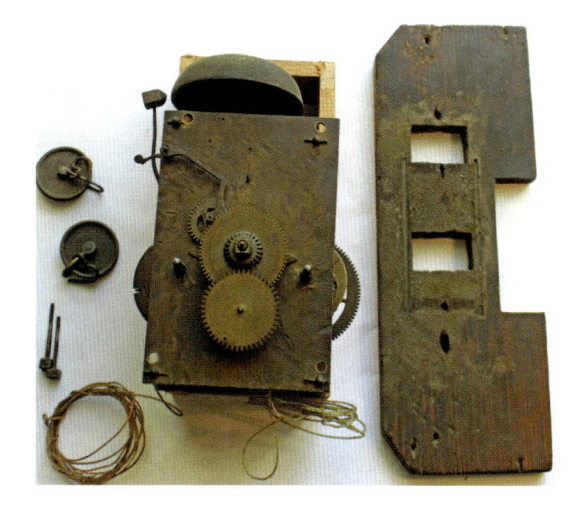

After removing the seat-board and tidying the gut-lines, remove the strike lifting piece, motion-work and any wheels associated with date work. (The layout of this early eight-day count-wheel clock differs in detail from a rack-strike layout.)

hour bridge in place but it is often accurately located by the addition of two steady pins. Remove the screws and gently pry the bridge loose from the steady pins.

The reverse wheel is sometimes held by a transverse pin and sometimes by a nut, but occasionally is retained only by the lifting piece. With the hour bridge removed, the minute wheel and the reverse minute wheel are slid off their centre arbor and post, respectively. The date wheel and moon-phase motion-work are removed at this stage, noting the positions of any pins relative to the hour wheel. (The wheels are meshed so that the date will change between midnight and four in the morning.) Finally, the bow spring, which provides the friction clutching arrangement, is removed from its shoulder on the centre arbor. (Note that the curve in the spring is arranged so that the outer ends point away from, not towards, the front-plate.)

Count-Wheel Striking, Eight-Day Plated Movements

Step 1 Remove Lifting Piece

The lifting piece is actuated by a pin on the minute reversing wheel; it is pinned to an arbor

Two 30-hour layouts: in the four-wheel or centre-wheel clock on the left, the motion-work is driven off the centre-wheel arbor, as in an eight-day clock (notice the seconds arbor protruding through the movement plate). The more common three-wheel clock on the right drives the motion-work through an extension of the main wheel arbor. The hour wheel, which has been removed to show the motion-work, is mounted concentric with, and on the same post as, the minute wheel and driven from the smaller front wheel on the left.

that carries the warning vane and lifts the hoop wheel detent and the count-wheel detent. Removal of the retaining pin will allow the lifting piece and the reverse minute wheel to be slid forward and removed.

Step 2 Remove the Date Intermediate Wheel and Hour Wheel
Sliding the date intermediate wheel and hour wheel away from the front of the movement will reveal the minute wheel and the hour bridge.

Step 3 Remove the Hour Bridge
As with the rack-strike movement, carefully unscrew the two retaining screws that hold the bridge on to the front-plate, taking care that because the steady-pins continue to fix the bridge to the plate, the bridge should be separated gently and evenly from the plate before lifting it clear. Removing the minute wheel will reveal the bow-spring located on the shoulder of the centre wheel arbor. The bow spring should be lifted clear of the centre wheel arbor.

Step 4 Remove Posts
The posts that support the lifting piece and the date intermediate wheel should be unscrewed gently from the plate, only if complete dismantling is required. They are unscrewed from the plate by gripping the square shoulder tightly and turning. As with all aspects of longcase clocks, the exception tends to be the rule and care should be taken just in case the posts are riveted to the plate.

Single-Weight (30-Hour) Plated Movements
Four-wheel, single-weight clocks are quite different in their wheel layout from the more common three-wheel type, and use a similar centre wheel/motion-work arrangement to ordinary eight-day movements, as described above.

However, the vast majority of single-weight clocks have three wheels in the going train with the drive taken through the front-plate by an extension of the main wheel arbor. The extension of the great wheel arbor supports a pinion, which drives the motion-work.

Step 1 Remove Lifting/Warning Piece
It may not be necessary to remove the lifting piece first – layouts vary and, in many cases, the pin that actuates the lifting piece is positioned at the back of the minute wheel. However, removal from the post is by removal of the transverse retaining pin or by unscrewing a small nut threaded onto the outer end of the post.

Step 2 Remove the Motion-Work
There is no intermediate reversing wheel in an ordinary 30-hour movement; instead the drive to the (clockwise) minute and hour hands is

directly from two wheels mounted on the (anti-clockwise) great-wheel arbor via a bow spring.

Birdcage Movements

Almost all birdcage movements are of approximately 30-hour duration between windings. Some, especially earlier country examples, are of a single-hand design, while others may have been made with minute hands or, alternatively, may have been single-hand clocks that were adapted at some time by addition of a minute hand with its associated motion-work. It is usually possible to detect the adapted single-hand clocks by the absence of minute marks on the outer edge of the chapter ring and only the half, quarter-hour marks on the inner edge.

Apart from the differences between one- and two-handed types, the layout of the motion-work and peripherals in birdcage movements depends largely on the internal wheel work. Where the going train has three wheels, the motion-work is more or less the same as the plated version and likewise, with four wheels in the going train, there is likely to be a bridge and minute reversing wheel, as in a four-wheel, single-weight clock or an ordinary eight-day.

The motion-work of a three-wheel posted (birdcage) movement is basically the same as in the equivalent plated movement. This two-handed movement has the hour wheel removed for clarity; in single-hand clocks there is just a smaller lower wheel on the main wheel arbor driving the hour wheel.

INSPECTING THE MOVEMENT

Once the back-cock, escapement pallet arbor and the motion- and strike-work have all been removed, it is usually best to inspect the internal wheel-work prior to further dismantling.

Depending on the result of that inspection, a decision may be made about whether or not it is necessary to separate the plates (i.e. whether complete dismantling is actually necessary). The reasons for dismantling are:

- Signs of excessive wear, which, if left uncorrected, will lead to greater problems in the future.
- Accumulation of dirt and dust, which, if left, will render the clock unreliable and cause undue wear.
- Dried out oil and consequent loss of lubrication in the bushing holes, which indicates the likelihood of impending excessive wear of both bushing holes and pivots (and, consequently, through incorrect meshing, the gear teeth).

Once the peripheral parts (the motion-work and strike-work) are removed, an un-obscured inspection of the two wheel trains is possible.

With the movement on its back on the workbench:

- Feel the fit of the pivots in their bushes. Clockmakers use the term 'shake' to describe the sideways movement of the pivot. (End-shake means the amount of movement in the direction of the wheel axis). The wheels are accessible and may be gripped between the fingers or, alternatively, use a pivot hook to push and pull the ends of the arbors to test for shake. Be aware that wear in bushes is often oval, so the shake in each pivot should be checked with this in mind.

- Look at the overall state of the plates, wheels, pinions and arbors – they should be clean and free from corrosion, rust, dust, grease and grime. Pronounced wear tracks often develop in steel pinions, especially if a movement has been heavily oiled, as abrasive dust sticks on the surface of the brass teeth and wears away the steel. If left uncorrected, further damage is inevitable.
- Look carefully at the ends of the pivots and their bushing holes. The bushing holes are usually countersunk from the outside of the plate to form an oil sink – a hemispherical depression designed to hold a tiny drop of oil that lubricates the pivot. After prolonged periods, the oil tends to dry out but it also mixes with brass rubbed from the bushings, turning the once fluid oil to a hard, green deposit. It is quite likely that a significant deposit of green-stained, dried oil is associated with serious wear in the bushes. (The green colour results from the oxidation of fine brass swarf rubbed from the bushing.)

Turn the movement over onto its front face, resting it on wood blocks to accommodate the length of the projecting arbors (the centre wheel arbor being much longer than the pallet wheel arbor) and repeat the detailed inspection process.

There is no easy formula for determining whether or not the shake in a pivot is excessive, but by inspecting the end of the pivot closely with a loupe or magnifying glass while testing the movement with the fingers, it is relatively easy to see if the shake is large relative to the diameter of the pivot. A good rule is any shake greater than about one-quarter of the pivot diameter is excessive. (Model engineers, who are used to working within fine tolerances, are often surprised by the quite large but perfectly normal amount of slack or shake in a longcase clock movement.)

DISMANTLING AND INSPECTING THE WHEEL-WORK

For the inexperienced or novice clock-mender, there is often the perception of a very large number of complicated wheels and arbors, and without carefully noting the exact position of every single part, re-assembly is quite impossible.

Not so; first, there are surprisingly few parts between the plates of a clock movement and, second, every part is different. Once the function of each part is understood, then the whole arrangement suddenly appears to be quite simple – astonishingly clever and elegantly designed but simple.

Plated Movements

Step 1 Unfasten the Front Movement Plate

Whether single-weight or eight-day, the procedure is similar – with the movement laid on its back on a frame on the bench, the pinned tops of the four pillars will be uppermost (sometimes movements were fitted with five pillars and, rarely, six and sometimes they are closed with latches not pins, but the principle is the same). Withdraw the pins by twisting and pulling with pliers or, if they are reluctant, use a long punch and a light hammer to knock them out from the inside of the movement.

The tapered pin is visible between the tips of the pliers. Place one tip against the thin end of the pin and the other against the side of the pillar to squeeze the pin from the hole.

Gently lift the front movement plate.

Step 2 Remove the Front-plate to Reveal the Movement

Gently lift the front-plate clear of the pillars and arbors, one corner first, leaving the wheel-work standing in the back plate, if possible. If the wheels fall into a jumble of parts as the front-plate is lifted, it makes no difference, but it might suggest very slack bushes.

Step 3 Record the Positions of Strike-Work Parts

In single-weight clocks, check and record the relative rotational positions of the hammer trip to the pin wheel and the warning and locking arrangements – a camera is very useful. Once the parts have been recorded, lift out the three arbors of the strike-work, including that of the hammer, and then the wheel-work, one train at a time. Keeping the two wheel-trains separate is not absolutely vital because no parts are interchangeable, but it is good practice.

The rope or chain may be left intact ready for re-assembly or cut and re-made when it is re-fitted. If the splice in the rope is suspect, then take the opportunity and cut it so that it doesn't get overlooked later.

Traditionally, clock-menders keep the two wheel-trains separate and quite often the great wheels or drums are marked accordingly. It is only the drums that are interchangeable and the

If possible, keep the wheel-work located on the back movement plate. It will now be possible to remove drive chains or ropes from sprocket wheels or gut-line from drums.

letters 'W' and 'C' are quite commonly scribed onto them, standing for watch and clock, what today menders call the going and strike sides. It isn't really necessary to divide the wheels in this way because, in practice, there is never any confusion if the drums are not removed from their respective arbors – there is only one possible

Make a careful note of the positions of pins, strike-work and especially the hoop-wheel/detent.

position and function for each part; but separating into logical groups does make for more ordered inspection and cleaning.

Birdcage Movements

The two trains of wheels are positioned one behind the other with the strike train to the back; they are held in place by three plates: the middle plate holds the front pivots of the strike train and the back pivots of the going side; the arbors of the strike-works are supported between the front and back plates. The bushing plates are fixed into the frame by steel wedges.

Step 1 Decide Whether to Split the Rope or Chain

All birdcage movements are of the single-weight type so they are fitted with either a chain or a rope. Prior to removing the wheel-work, a decision should be made about whether or not to cut the rope on the splice and re-make it later during re-assembly or whether to keep the rope intact and work around it if the splice is sound. Where a chain is fitted, it is usually easier to open a chain link and unthread the chain from the two wheels first.

Step 2 Locate and Remove the Wedges that Hold the Movement Plates

Place the movement on its base on blocks on the work-bench and locate the wedge that holds the top of the back plate into the top plate of the frame; tap it upwards with a drift and a light hammer. The back plate, although located by two pegs in holes in the bottom plate, is now lifted up into the top plate, allowing room to lift the pegs clear of the bottom plate, so that the bottom of the plate can be swung clear.

Step 3 Record and Remove The Strike Train

Having recorded any details about the relative positions of strike-works and having removed the back post, the wheels and arbors of the strike train and the strike-work arbors may be lifted clear and set aside.

Step 4 Remove Motion-Work and Front Movement Plate

After removing any motion-work, locate the wedge that holds the top of the front post, tap it upwards with a long flat-ended punch and a light hammer. Fiddle the front post out by first lifting the top end clear of the top plate and then lifting the pegged end from the bottom plate.

Step 5 Remove the Going Train

Lift the three (or occasionally four) wheels of the going train clear of the middle post and set aside.

Step 6 Remove the Middle Post

Remove the wedge holding the middle post and remove that post from the frame. It is not necessary to mark the middle plate for front and back alignment, but they are nevertheless sometimes marked and care should be taken to avoid confusing the two sets of bushing holes when inspecting.

The two trains of a birdcage movement may be dismantled separately. The vertical bushing plates are provided with pegs that lock into holes in the bottom plate and steel or brass cotters (wedges) to hold them firmly in the top plate. The cotters are removed to enable extraction of the bushing plate and wheel train.

CLEANING

De-Greasing

Occasionally, clock movements are so coated with oily, greasy or even tarry deposits that the best first step is to use an engine de-greasing fluid obtainable from car accessory shops. The small parts may be soaked in the fluid and it may be painted onto larger surfaces such as movement plates. Subsequent washing in warm water results in oil-free surfaces from which corrosion and tarnishing may be removed by immersion in a cleaning solution followed by polishing.

Hand Cleaning Versus Ultrasonic Cleaning Machines

For the professional clock-mender, an ultrasonic cleaning machine is a basic necessity in everyday use. The parts are placed into a wire basket, which is immersed in the heated cleaning fluid; the agitating effect of the ultrasonic device lifts the grime off the surface. A clock-mender may use his cleaning machine two or three times while overhauling a single clock movement: first, to clean prior to repair; then, after repair work, to remove metal swarf and debris; and, finally, after polishing to remove finger prints, residues and debris from polishing, especially from crevices and between the teeth of wheels and pinions.

Ultrasonic cleaning is not an absolute necessity and for the amateur on a budget, hand cleaning usually means immersing the movement parts into a bowl of cleaning solution and then polishing with a metal polish to remove stubborn tarnish.

For the amateur mender with limited resources, immersion in warmed cleaning fluid is workable but requires more polishing after. (While the ultrasonic machine tends to leave bright brass surfaces, the surface appearance after cleaning solution immersion is different to that after plain water immersion. Acidic solutions tend to leave a copper appearance on the metal, while alkaline solutions tend to leave a zinc appearance.)

After immersion cleaning, the metal surfaces will require careful polishing to remove residues without causing damage, especially to pin wheels and the like.

As an alternative to immersion, the cleaning solution may be made up into a gel by mixing in a little wall-paper paste and applying it to the tarnished parts.

After a final rinse with hot water, the parts are dried with an ordinary hair drier and set aside for repair or reassembly.

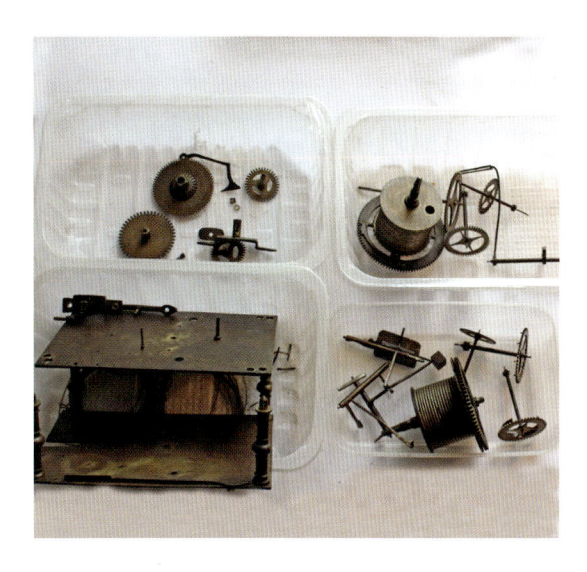

The dismantled movement is separated into groups ready for cleaning.

Cleaning Solutions

There are two approaches to cleaning solutions for tarnished or corroded metal (brass or steel) and either is suitable for simple immersion or for use in an ultrasonic cleaning machine.

The acid solution uses a weak acid, typically citric acid (lemon juice), mixed with a wetting agent such as liquid soap or ordinary washing-up liquid. The solution can be made up with a tablespoon of citric acid crystals in about two litres of hot water.

Alkali cleaning solution uses household ammonia mixed with a wetting agent. Household ammonia may be diluted with only warm water and even then the fumes are quite noxious.

Several commercial solutions are available from horological suppliers.

Polishing

Ultrasonic cleaning is usually quite thorough and hardly any further polishing is required. By comparison, simple immersion is less efficient and gives rise to a loose metallic coating on the surface of brass.

Brass polish, which is widely available as abrasive creams or pastes (Brasso, Peek, etc.), is applied very thinly and, in the case of heavy tarnishing, it may be applied with extra fine wire wool rather than a cloth. It is essential to polish small areas at a time, working around collets and pins with a small, fine brush. From a conservation perspective, less polishing is preferable – just enough to leave a flat, clean metal surface without unnecessarily removing metal.

Steel parts may be polished with chrome polishes (available from motor factors), which are more effective at removing surface corrosion marks than brass polishes.

When metal polishes are used, the polished parts should be cleaned very thoroughly in hot soapy water to ensure that all traces of abrasive are removed, especially from between the teeth of wheels and pinions.

The importance of thorough rinsing and equally thorough drying cannot be over-emphasized – the good results of cleaning are often negated by allowing moisture and potentially corrosive cleaning agents to remain on brass and steel surfaces.

The results of thorough examination are often surprising. This badly worn centre-wheel back pivot is on the point of complete failure.

The cleaned movement is part re-assembled to systematically check for wear and faults.

FURTHER INSPECTIONS AFTER CLEANING

Once the movement parts have been cleaned, they may be inspected closely for signs of wear.

Pivots

Pivots should be examined first under a glass to establish whether or not any work is required: they should appear parallel, smooth, evenly polished and definitely devoid of ridges or rough surface. Damaged pivots will be apparent and should be rectified before the damage spreads.

Bushes

The fit of the pivots should have been checked already prior to removing the wheel-work and any noticeable shake noted for attention. Each bushing hole should now be inspected under a glass to check any ovality and the soundness of any earlier repair work. (Occasionally, when a clock is re-bushed, the new bush works loose and the complete loss of a replacement bush may cause serious damage.) An oil sink should be present, countersunk into the outside end of the bush. If it has been omitted during the replacement of a bush, it should be dealt with prior to re-assembly.

Wheels and Pinions

The teeth of the brass wheels are intended to roll on the surfaces of the leaves of the steel pinions. Wear in bushes and pivots causes misalignment that exacerbates the wear in the teeth because it brings about a rubbing action between the teeth. The steel leaves of the pinions tend to wear more and they should be inspected closely. If the wear is excessive, repair work may mean simply moving the wheel slightly on the arbor to present new mating surfaces on sets of teeth.

Wheels should also be checked for lost teeth or precarious historical repair work to previously lost teeth.

Pins

Pins are used on the strike train wheels; to trip the hammer, they are fitted to the pin wheel in an eight-day movement and the great wheel in a single-weight clock. In addition, they are found on the warning wheel in all clock types and quite often in single-weight clocks, as an alternative to the hoop wheel, to lock the strike train. The pin wheel is particularly prone to wear because of the rubbing action of the hammer-lifting piece and the pressure from the hammer spring.

If the wear is noticeable, just rotating the pins to present a new rubbing surface may be sufficient. If the wear is excessive, the part should be set aside for repair, which will entail removing the damaged pins and replacement with new steel pivot wire of the same diameter.

Escapement Pallets

The wearing surfaces of the escapement pallets should be inspected carefully. After thirty-one and a half millions ticks every year it is natural enough that over the space of many decades or centuries, the faces of the pallets eventually develop wear tracks.

Inspecting the Strike-Works

One aspect of count-wheel striking movements that is frequently overlooked is the degree of

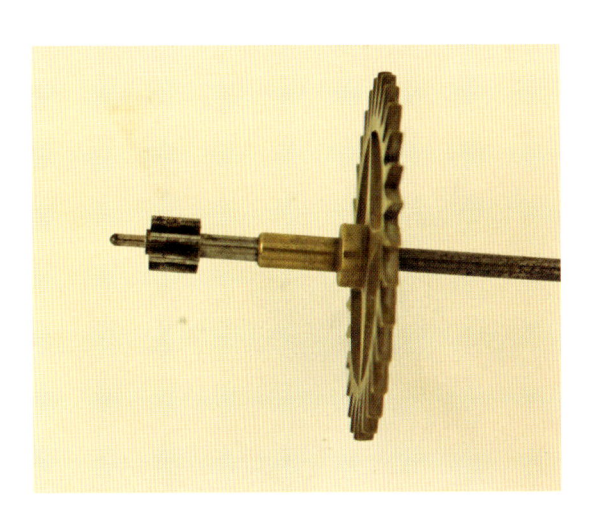

Wear in pinions is often a symptom of abrasion in the wheel-work from dust sticking to an oily film. This pinion is just beginning to show wear tracks, which alter the meshing geometry of the wheel and pinion.

The rubbing surfaces of the escapement pallets are prone to wear, and once wear tracks have developed, the escapement geometry is upset. Pallets have often been ground back to a flat surface and sometimes re-faced, so it is always worth checking the overall geometry by accurate measurement and drawing.

wear in the strike-work bushes. Count-wheel striking is simple and efficient but a surprisingly small amount of wear in the bushes of the warning and detent arbors is apt to make the chiming erratic.

RE-ASSEMBLING THE MOVEMENT

Prior to re-assembly, the bushes should be pegged to remove any deleterious material. Softwood pegs are twisted in the bushes to ensure that they are absolutely clean. The peg-wood is trimmed to a fine taper and then twisted in the bushing hole to pick up any grime or trapped polish. The process is repeated until when the peg-wood is withdrawn from the bush it is absolutely clean.

When the maker of a longcase clock movement originally completed his work by assembling his clock movement, it undoubtedly worked perfectly. Re-assembling a dismantled clock with all the parts in their correct order will repeat that original assembly exactly, and the clock will work perfectly as it did on that first day, as long as the wear is not so great that the resulting increase in internal friction does not exceed the driving force.

Plated Movements

Step 1 Fit the Peripherals
If they are fitted to the inside face of the back-plate, screw the hammer spring and the bell-stand into place.

Step 2 Place the Back Movement Plate onto a Firm Surface
The back-plate is placed in a movement stand, or on a wooden frame, or even on two blocks on a clean surface on the work bench, with the inner face upwards. (Pivots often extend beyond the back face of the movement plate.) Then, taking each train in turn, the back pivot of each wheel is placed into its respective bush. (Except the escapement arbor, which is fitted later with the back-cock.)

Contrary to some novices' perceptions, there are usually only nine wheel arbors and the hammer arbor to fit into an eight-day movement.

Although a 30-hour movement may have one wheel arbor less, it does have two extra strike-work arbors.

Step 3 Work the Front-Plate onto the Pivots
The front-plate is lowered carefully onto the assembly of upstanding arbors – carefully, because the pressure on the pivots should not mark the inside of the front-plate and the pivots themselves, being hardened steel, are susceptible to breakage.

It is usually easier to place the assembling movement onto a stand to bring it to eye-level, so that the position of each pivot can be watched and adjusted to line up with its respective bushing hole.

Eventually, and with patience, the pillar holes in the front-plate will naturally lower onto their respective pillars, closing properly onto the shoulders.

Once this has been achieved, the transverse tapered steel locking pins can be temporarily pressed into place to hold the plates together.

Step 4 Check the Strike-Work
After checking that each of the wheel trains turns easily with finger pressure applied to the great wheels, the relationship of the strike-

With patience, and a light touch, the front movement plate will let itself onto the pivots progressively under its own weight.

Clock Maintenance

work should be checked. In all types of movement, the hammer trip should be clear of the pins and, consequently, the hammer spring relaxed when the strike is at rest with the train locked. In addition, the warning wheel should be able to rotate sufficiently for the strike cycle to release cleanly and for the warning pin to rotate onto the warning piece. That action may be checked by temporarily placing the lifting/warning piece onto its post and watching the action. The warning wheel will be correctly oriented if the pin is opposite the warning vane when the strike is locked.

In 30-hour clocks with a hoop wheel, the locking piece should be just about ready to drop into the slot of the hoop as the hammer trips. Where a pin wheel is used instead of a hoop wheel, the locking pin should be just slightly in advance of the locking piece as the hammer strikes. In either case, it is the detent dropping into the slot of the hoop wheel, or against the locking pin, that locks the train, rather than the count-wheel detent dropping into its slot. So temporarily mounting the count-wheel with its driving pinion and the count-wheel detent, will indicate if the set-up is correct.

Step 5 Prepare the Assembled Movement for Oiling
With the internal strike-work set up correctly and the tapered steel pins pressed firmly into place to hold the front-plate, the escapement arbor and back-cock are re-fitted and the movement is now ready for oiling.

Using good quality clock oil, dip a thin wire oiler into the oil and apply a small drop of oil to each of the pivots and to the pallets of the escapement. If too much oil is applied, the sink will fill to overflowing and, as the excess runs down the plate, it will draw the rest of the oil away from the pivot. With the exception of the escape wheel teeth, care should be taken to avoid any oil contamination of the wheel teeth.

Step 6 Re-Fit Motion-Work and Strike-Work
The motion-work and strike-work are re-fitted to the front-plate taking care to re-fit the transverse retaining pins. In rack-striking movements, the rack spring should engage with the drop of the rack so that it applies pressure to rotate the rack, so that the rack tail makes contact with the snail.

In all movements, apart from old single-hand types, the relationship of the reverse minute wheel of the motion-work to the lifting piece and the minute hand square is critical: the mesh of the reverse minute wheel with the cannon

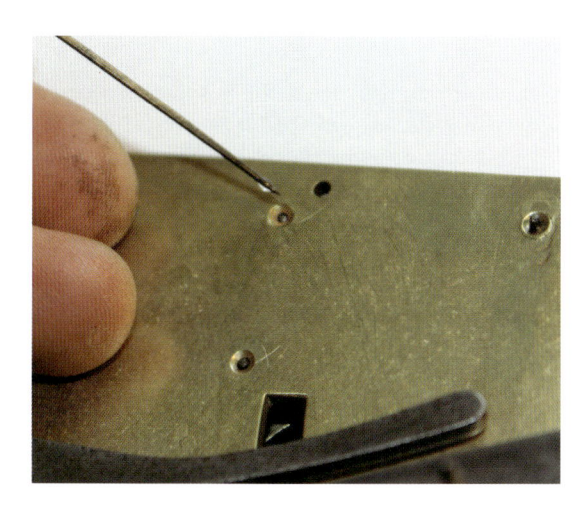

Only sufficient oil is needed to fill the gap between pivot and bushing.

Setting up the correct meshing of the reverse minute wheel. The minute hand is approaching the hour and the pin on the reverse minute wheel has lifted the lifting piece. The strike train is on warning and, if the meshing is correct between the two wheels shown, the lifting piece will drop off the pin exactly as the minute hand reaches the XII position to start the strike cycle.

wheel should be adjusted so that, as the lift drops off the pin, the minute hand is pointing directly to twelve.

In rack-striking movements, when the warning trips, the pin on the rack-tail should fall onto the middle of the step on the snail, away from the riser.

Step 7 Re-Fit the Dial and Hands
With the motion-work set up correctly, the dial is re-fitted using tapered steel pins. The hands are re-fitted: first, the hour hand is attached, usually by means of a small screw into the shoulder of the mounting tube; second, the minute hand is placed in its correct orientation onto its square and the washer must be pressed back to tension the bow spring before the pin can be slipped through the hole in the end of the centre arbor.

SETTING UP THE CLOCK

Once the maintenance work is complete and the movement and dial re-assembled, it only remains to replace the movement into the case, set the beat of the escapement evenly and check that the strike is working correctly. It is the escape wheel and its interaction with the pallets that are the source of the 'tick'.

Step 1 Attach Pendulum and Weights
Having secured the movement to the seat-board and positioned the seat-board into the case, the usual procedure is to hang the pendulum first and then attach the weight(s) to the pulley hook(s), but some clock-menders prefer to hang the weights first to listen to the beat of the unregulated escapement, in order to make sure there is no catch in the train and everything is free.

Step 2 Check Pendulum Locations of Escapement
Look carefully at the two parts of the escapement and how they inter-relate and make sure that the escapement looks roughly level with the pendulum hanging at rest. One of the escape wheel teeth will be in contact with one of the pallets. Holding the pendulum between finger and thumb it is swung in the direction that releases the contacting pallet. As soon as the pallet releases, the characteristic 'tick' will be heard as the opposite pallet will stop the escape wheel.

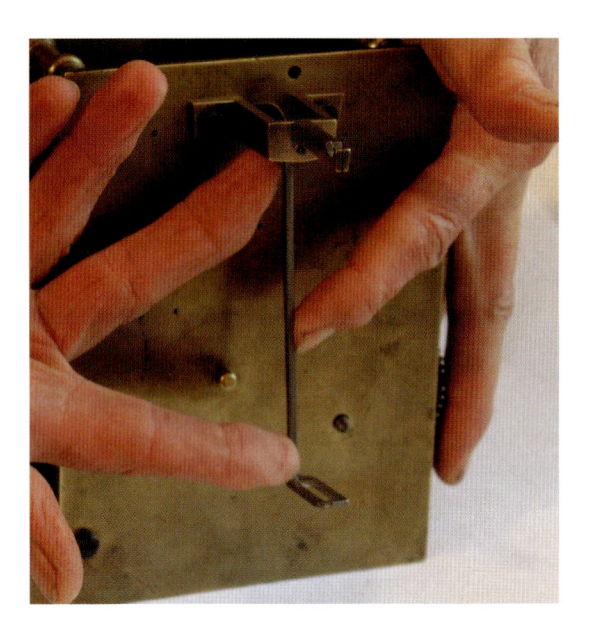

The crutch stem is annealed to make it malleable: forcing a slight bend into it at about the midpoint will adjust the beat of the clock.

Step 3 Set the Clock on Beat
Repeat the above step a couple of times to ensure that consecutive ticks are within reasonable swings of the pendulum. Then release the pendulum from at least the displacement just now established. Leave the clock to settle down for a few minutes and then start to listen to consecutive ticks, which should be equally spaced.

There are 'engineering' approaches to beat-setting that rely on noting the relationship of the at rest position of the pendulum to the extremes of the swing required to trip the 'tick', but, in general, clock-menders tend to rely on their senses, primarily hearing but also sight.

First, by listening to the tick with the trunk door open, decide which side the tick is biased towards. Then carefully bend the crutch so as to bend the forks towards that side. Listen again and keep repeating the process until the tick sounds evenly spaced.

Step 4 Regulate the Time-Keeping with the Pendulum Rating Nut

With the clock running on beat, the last job is to regulate the time-keeping. The regulating nut adjusts the position of the pendulum bob in a fine or precise way. It is possible to calculate the number of turns or part turns of the regulating nut using the fundamental formula for simple pendulums but, as a rough guide, 1mm of movement of the regulating nut is equivalent to an adjustment in time-keeping of 43sec per day.

Accurate regulating may take a week or two and, for eight-day clocks, the regulation should recognize the special feature of drum-wound clocks, which tend to run backwards during winding.

DERIVATION OF THE REGULATION FACTOR

The period of swing of a simple pendulum is governed by the natural law:

$$T = 2 \times \pi \times \sqrt{\frac{L}{1000 \times G}}$$

Where T (sec) is the period of the complete oscillation (i.e. T would be 2sec for a normal longcase clock and 1sec for a pendulum); π is the circle constant (the ratio of circumference to diameter for any circle); L (mm) is the length of the pendulum; and G (m/sec/sec) is acceleration due to gravity.

A small increase (l) in the length gives rise to a small increase (t) in the period of swing from which for small values of t simplifies to:

$$l = 2 \times T \times t \times G \div (4 \times \pi^2)$$

or, for a normal longcase pendulum, when $T = 2$:

$$l = t \times 9.81 \div \pi^2$$

It is easier to work with daily time error in seconds rather than the percentage of each second (for example, with 15 seconds error per day than 0.0173611 per cent), so the pendulum correction measured in millimetres is approximately equal to the daily error in seconds multiplied by 0.0115 (pendulum bob goes up for faster, down for lower).

For example, if the clock is fast by 4sec per day, the correction is 4×0.0115 or 0.046mm.

If the pitch of the regulation nut thread is half a millimetre, the adjustment will be approximately one-tenth of a revolution. Clearly, for accurate control over regulation, the regulating nut thread should be as fine as possible.

FINAL TESTING

Finally, with the clock assembled but without the dial, it is set up on a test stand that gives good visibility so that any faults that develop may be examined.

The clock will only run properly with the hands fitted, otherwise there is no pressure on the bow spring to drive the motion-work.

A simple test stand allows good visibility and easy access.

Chapter 5

The Movement – Some Simple Workshop Procedures

Complete dismantling is a pre-requisite of almost all restoration work to longcase clock movements. The exceptions, which are dealt with in the first parts of this chapter, concern the pendulum and its suspension, the weights, pulleys and chains and ropes, and then the parts that are mounted between the dial and the movement proper, including the strike-work.

The remainder of the chapter deals with the two most common ways of correcting the effects of wear: first, the procedure for dealing with gradual roughening and wearing away of pivots; and, second, the related wear enlargement of the movement plate holes (the bushes) that support the pivots.

We have included the pivot and bushing topics in this chapter about basic procedures because, although the work may seem a little tricky to the novice, they are probably the two most fundamental and most frequently required corrections in restoring a clock movement to an acceptable level of mechanical efficiency. As wear increases over the years, the friction losses in the bushes increase but, more significantly, the wheels and pinions tend to undergo changes to their meshing geometry, with a further increase in internal friction in the gear train. Restoration work corrects the slack or shake in the bushings and, by correcting the drift in the relative positions of arbors, returns the movement to a state where it can tick freely without likelihood of being stopped by friction losses. No amount of cleaning and oiling can properly rectify pivot and bushing wear, and in fact an excessive gap between pivot and bushing will render oil useless because the contact between the two rubbing parts is concentrated at one line contact.

REPLACING A BROKEN OR DAMAGED PENDULUM SUSPENSION SPRING

The suspension spring or, in some clock-repairers' jargon, the feather, supports the weight of the pendulum with enough flexure to allow it to swing easily. Over the quite small amplitudes of swing required for anchor (and deadbeat) escapements, the suspension spring is fairly immune to the effects of repeated bending, and it is most likely to sustain accidental damage or breakage when it has been removed from the clock, and especially during transport. The development of a sharp kink is an obvious flaw, which is liable to develop into a break, but in any case a kink tends to set up a wobble in the motion of the pendulum.

Step 1 Decide What to Replace

Replacement suspension springs are readily available from horological suppliers. They come in various lengths, depending on the manufacturer,

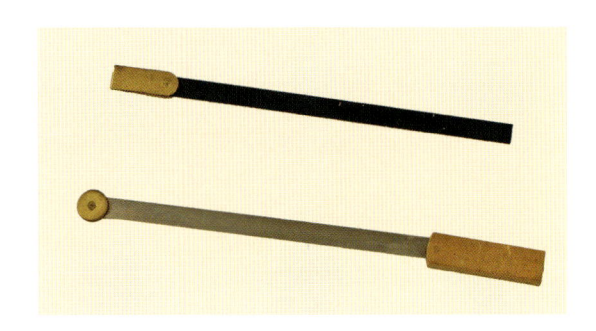

Types of off-the-shelf suspension spring. As an alternative to using made-up spring blanks, it is possible to use light clock spring or any light spring steel.

Step 2 Removing the Parts to be Discarded

Remove the rating nut and pendulum bob before attempting any work on the top end of the pendulum rod, then proceed with removing the damaged parts:

- **Wire pendulum** – gripping the rod in the soft chops of a bench vice, unscrew the old block that contains the damaged suspension spring.
- **Strap pendulum** – carefully file the end of the pin that holds the spring into the crutch block and then tap it out using a small flat-ended punch, and then pull the old spring from the slit in the block. If the crutch block is badly worn, out of alignment or just too thin, it should be replaced with a suitable brass block pinned to the top end of the strap.

Step 3 Fitting the New Spring

Wire Pendulum, New Crutch Block If the new suspension spring assembly is correct for length and fit to the crutch fork, the type of new suspension spring pre-fitted with a crutch block may be the best option. It may be screwed directly to the threaded end of the pendulum rod but, in most instances, the mismatch of threads must first be corrected by re-cutting one or other.

Strap Pendulum (Retaining the Old Crutch Block) Measure and cut the new spring and mark the location of the pin hole. The pin hole is made by punching with a flat-ended punch of the right diameter onto a lead block. Do not use a sharp punch to pierce the spring as it will split the spring rather than punching out a clean hole. Alternatively, make a small dent with a sharp punch, file off the corresponding bump on the other side and open the resulting hole with a broach. The holes in the brass block should be slightly countersunk so that the ends of the pin can be riveted over. The retaining pin may be steel or brass but must be a good tight fit and, once riveted over, it must be filed back flush to prevent snagging in the crutch forks. Care should be taken that when riveting the pin, the spring is not gripped solidly in the

A pendulum suspension spring is suspended from the back-cock and should be a sliding fit in the forks of the crutch.

and with either a single, riveted brass block from which it hangs, fitted into the notch in the top side of the back-cock chops, or with two blocks, the second being the crutch block at the bottom end, threaded usually at 3BA to take the pendulum rod.

The first (plain) type requires fitting to the original crutch block still attached to the pendulum, so it is more applicable to strap-type pendulum rods. The second type, pre-fitted with a new crutch-block will need to be fitted to the original crutch and is more applicable to the earlier style of wire pendulum rod.

The critical dimension is the distance from the point of suspension in the chops of the back-cock down to the crutch block. It is a simple matter to offer up a replacement suspension spring to check the length and also the fit of the crutch block in the fork of the crutch, which should be an easy sliding contact, but not loose, or otherwise the impetus from the crutch will be diminished and the clock will not run smoothly.

The threaded top of a pendulum rod (right); the brass block at the lower end of the suspension spring screws onto the rod (centre). The strap-type pendulum (right) is found in later clocks; the brass block is riveted to the steel strap.

block; it should be able to hinge around the retaining pin enough to correct its alignment under the weight of the bob when suspended in the clock.

REPLACING OR REPAIRING A DAMAGED OR BENT CRUTCH

The second most common cause of damage to longcase clock movements is accidentally bending or breaking the crutch during storage or transport. A clock movement removed from its case is prone to falling backwards onto the projecting crutch forks and, although the crutch is intended to accommodate the small amount of bending that is necessary to adjust the beat, severe bending or impact can result in a fracture. Repairs that involve joining broken parts with sleeves and the like are mostly unsightly and mechanically unsatisfactory; a replacement is usually the neatest and soundest approach.

Examining the detail of the remnant crutch and its collet is the first step to arriving at a repair plan, because there were several alternative ways that clockmakers used to attach the crutch to the escapement arbor.

The most common method of attachment is a hole drilled through a thickened part of the arbor to take the top end of the crutch rod. The top of the crutch rod is either soldered into the hole or the top riveted over. It is sometimes possible to drill out the old crutch from the arbor and re-fit the replacement crutch into the same hole. There are many variations on that method, often involving a widened section of the top of the crutch rod fitted into a groove in the underside of the arbor collet, designed to prevent rotation of the crutch rod. Alternatively, the top end of the crutch rod is enlarged to a circular plate, which is drilled to the size of the arbor and soft soldered to the brass collet.

In the event that a complete replacement of the crutch is necessary, replacement crutch blanks are obtainable from horological suppliers. They are stamped from steel but they are only blanks and require shaping and finishing prior to fitting to the escapement pallet arbor.

Step 1 Remove the Pallet Arbor

Remove the back-cock as described in Chapter 4 and withdraw the pallet arbor.

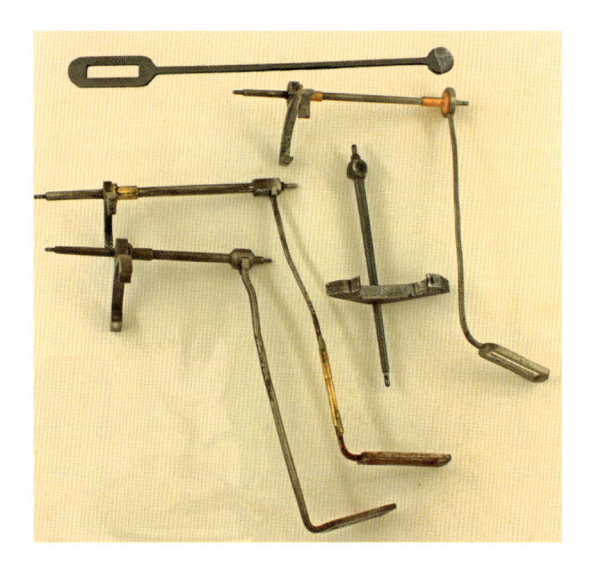

An off-the-shelf crutch blank (top) can be adapted to the two alternative methods of attachment to the escapement pallet arbor.

Step 2 Remove Broken Parts but Do Not Discard

Remove the old top end of the crutch by heating gently with a spirit lamp to melt the solder. It is inadvisable to use a gas lamp because the intense heat is difficult to control and is likely to affect the temper of the pivot and spread to the solder fixing of the pallet collet.

Step 3 Shape the Forks

Starting with a crutch blank, the gap between the forks should be adjusted for a sliding fit of the pendulum crutch block. The gap should be parallel and true, to keep the driving impulse from the clock even. Once the gap is correct, remove the excess steel from the outside profile of the forks to give a neater appearance. Finally, the forks are bent to form a right angle to the crutch stem.

Step 4 Fit New Crutch

Measure the diameter, centre punch, and drill the top end of the new crutch for interference fit to the arbor. A pillar drill will ensure that the hole is drilled perpendicular.

Clean up the arbor and re-tin, if necessary. The tinning should be the thinnest cover.

Tin the hole and the front face where it mates to the collet.

Assemble the parts so that the line of the crutch is, as close as possible, a right angle to a line between the pallets and then heat with a spirit lamp just until the solder runs.

Step 5 Re-Fit the Pallet Arbor

Return the pallet arbor by guiding the crutch through the hole in the back-plate, then locating the front pivot into its bushing and finally replacing the back-cock.

Once the movement is back in the clock, the beat will require setting by bending the crutch, as described in Chapter 4.

REPLACING BROKEN OR WORN LINES

Chains or ropes for single-weight clocks and gut or, occasionally, brass or steel wire, especially for longer duration clocks, are all prone to wear and ultimately breakage. All are obtainable from suppliers and, in general, a sound approach would be to replace like with like, except that natural gut (usually called cat-gut but coming from sheep) is the correct cord for ordinary eight-day clocks. Laid cords and synthetic fishing twines are unsatisfactory.

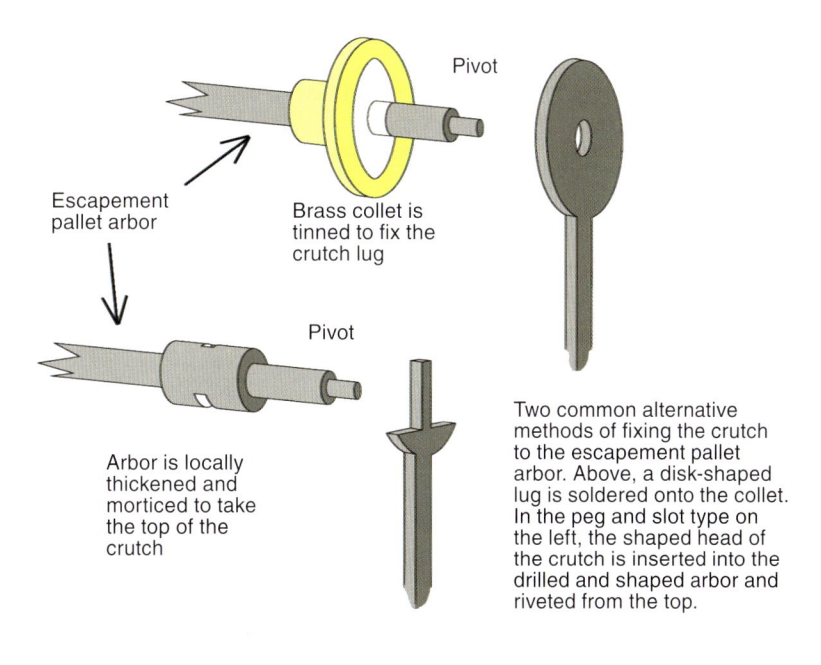

Schematic of the soldered collet and peg and slot types of crutch fixing.

100 The Movement – Some Simple Workshop Procedures

Conversion kits are available for altering 30-hour clocks from rope to chain drive (the spiked sprockets for rope drive are not compatible with chain). However, there are pros and cons – questions about pointlessly changing an historical clock carry some weight. The justification based on the excessive amounts of dust that enter the movement from rope drives is not necessarily valid if that dust is not particularly harmful. Rope-driven clocks are no more prone to wear than chain-driven.

Eight-Day Clocks

Step 1 Remove the Old Gut-Line
The return end of the gut-line passes through a hole in the seat-board and is tied to a peg. Remove the discarded line from the seat-board and then unwrap the turns of line from the drums to allow a sufficient length to be pushed into the hole in the drum to allow the knotted end in the drum to be hooked out and completely removed.

Step 2 Replace Drum End of Line
Bend a curl into the end of the gut-line and pass it through the hole in the tread of the drum,

Woven cord, natural cat gut (made from sheep's intestines) and chain are all available from specialist suppliers.

pointing the curl towards the access hole. Withdraw enough line through the access hole to form the knot. Moisten the end to be knotted, to make it supple enough to form a figure-of-eight knot.

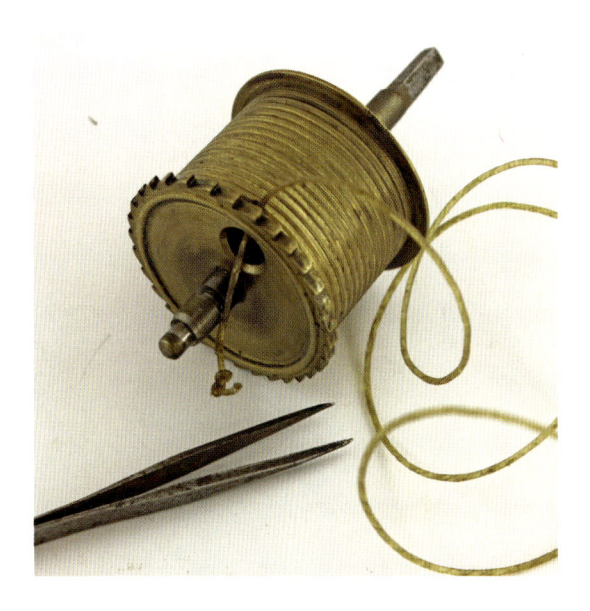

Replacing the Rope on a 30-Hour Clock

Clock rope consists of an outer sheath of about a dozen woven strands and an inner core of several straight strands. There are various ways of joining the rope ends and the only stipulation is that the join or splice should have the same diameter as the rest of the rope. Consequently, if the ends are to be overlapped to form a splice, some of the strands should be removed and, preferably, they should be trimmed out in an uneven or random way.

In each case it is essential to first ensure that the rope is threaded correctly through the clock, the pulley and the counter-weight.

Method 1
Cut the ends square and bind them tightly with button twine. Thread a large needle with button twine and, placing the two ends together, sew through the rope joining the ends firmly together.

The end of the gut-line is fed through the small hole in the tread of the drum and drawn from the hole in the cheek in order to tie the knot.

Method 2

Push back at least 70mm of the outer part of each end of the rope to expose the core. Divide the inner core into its three or four strands, knotting opposite pairs so that the knots are spaced along the join. Take the outer sheath of one side and, after removing every second strand, arrange the remainder over the knotted inner core and bind with button twine. Grasp the bound ends with the knotted core between the points of large tweezers and push as far as possible into the outer sheath of the other rope. The strands should now overlap by about 50mm. Remove every second strand from the woven outer sheath by snipping farthest from the join and pulling the cut strand through the weave. Finally, bind the join by a combination of stitching through and binding around with button twine. The best results are obtained by staggering the cut ends as much as possible.

A step in the centre arbor slightly protrudes from the front of the movement plate. It supports the bow spring without letting it rub on the movement plate.

MOTION-WORK FAULTS

Loose Reverse Minute Wheel and Other Posts

Several posts are mounted onto the front movement plate, including the one that supports the reverse minute wheel. Posts are steel with a thread (usually about 3BA) to fix them to the plate. They are almost always made with a slight taper and some means is provided to secure the part they are intended to support (i.e. the reverse minute wheel in this case, usually the end is drilled to receive a tapered pin). The threads of posts occasionally become loose, which causes meshing problems in the motion-work. Once the thread is slack, continued tightening is likely to completely strip the remaining thread, so extreme care should be taken to avoid the need to turn a new post.

As a conservation matter, an application of a thread-locking fluid to prevent further movement of the post may be sufficient and far more acceptable than making material changes to the clock.

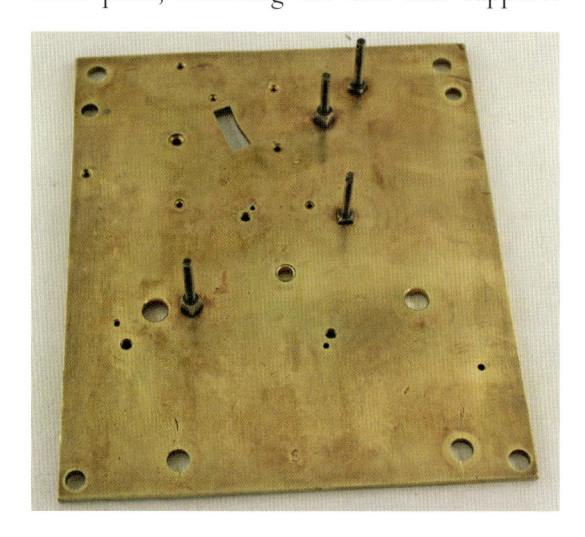

Posts fitted to the front movement plate of an eight-day clock. They differ in length, diameter and offset from the plate (i.e. the height of the nut). From front to back they support rack, reverse minute wheel, rack hook and lifting piece.

Broken, Weak or Inverted Friction Bow Spring

The bow spring is the slipping clutch that allows the time to be adjusted by letting the hands slip relative to the wheel-work. During normal operation of the movement, the bow spring transfers torque from the centre wheel arbor to the minute wheel.

The spring is located behind the minute wheel in a centre-wheel clock (i.e. an eight-day clock or a four-wheel, 30-hour clock) or in an ordinary 30-hour clock, on the front extension of the main wheel arbor behind the wheel that drives the motion-work.

The symptom of a broken or weak spring is a tendency for the hands to slip (often at about twenty to eight) and it is often misconstrued as a sign of a very serious fault. The diagnosis is made by removing the pendulum and weights and applying pressure to turn the minute hand clockwise. In a healthy clock, although the hands slip past the wheel train, they should not turn too easily and there should still be enough friction to drive the escapement. So, winding the minute hand forward, with neither weights nor pendulum attached, should give rise to a ticking noise. (Warning – does not apply to deadbeat escapements.)

One of the common faults in bow springs happens when it is incorrectly re-assembled facing backwards and, consequently, is unable to exert a pressure on the back face of the minute wheel.

Step 1 Prepare the Movement
Remove the movement from the case, and the hands and dial from the movement.

When fitted, the bow spring should sit onto the step on the centre arbor, clear of the movement plate with the outer ends curled forwards.

The hour-wheel bridge may look symmetrical but they usually only fit one way.

Step 2 Remove the Hour Wheel to Gain Access to the Hour Wheel Bridge
Remove the hour wheel from the bridge pipe to expose the screws that fix the hour bridge to the movement plate. It may be necessary to remove the rack together with the hour wheel if there is insufficient clearance.

Step 3 Access the Spring and Inspect
Unscrew the bridge screws and remove the bridge and minute wheel to expose the friction bow spring. It is possible that a previous repair has resulted in the spring being re-assembled back-to-front; the outer ends of the spring are bent forward and should bear on the back face of the minute wheel. In that case, just reverse the spring and re-assemble, otherwise proceed to replace the spring.

Step 4 Make Up a New Bow Spring
Mark out and cut a piece of 15 thou (0.3mm) brass shim plate and drill the centre for the small diameter of the centre wheel arbor. (There is a step in the centre wheel arbor upon which the spring seats; the spring should not bear against the movement plate.)

Step 5 Temper to Spring Hard
On a flat, anvil plate, hammer the piece repeatedly with a light hammer, testing it for spring

periodically. Hammering on only one side will induce a curl into the new spring and, after a short while, it will be springy enough. Finish by slightly turning the outer ends back and fit to the centre wheel arbor.

Hour Wheel Bridge Problems

When a clock has been re-assembled and is reluctant to run smoothly, the fault usually lies in some odd source of friction that results from the detail of re-assembly. The hour wheel bridge is one of those candidate parts, which, when inadvertently rotated through 180 degrees, is prone to rubbing the minute wheel pipe.

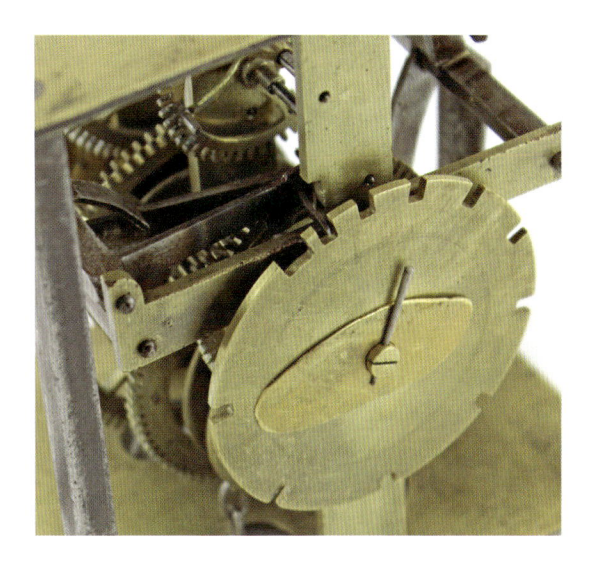

STRIKING FAULTS

Some strike faults are common to both rack and count-wheel systems and some are peculiar to the type. Rack-strike faults tend to be associated with the interaction of the rack with the hour wheel snail, including the operation of the rack spring and the gathering pallet.

Count-wheel striking is intrinsically more robust but equally prone to wear. Erratic strik-

Although the system of using a warning device is later, this eighteenth-century birdcage movement would have been recognizable to a clockmaker of the fourteenth century. The count-wheel detent or latch has dropped into the slot after striking two o'clock. It should not rub against the sides of the slot.

ing in count-wheel clocks, which are almost always 30-hour movements, can often be traced back to shake in the bushes of the detent arbors, which tends to cause imprecise location of the latches into both the hoop and count-wheels.

A detailed understanding of the nature and detail of the strike-work is essential for understanding the nature of any faults and planning the repair work. In general, the problem is often exacerbated by a legacy of previous attempts at repair that involved bending or filing.

When the strike becomes erratic, very careful examination on a test stand, with the dial removed, is the first step and usually the problem will become obvious; for example, if the count-wheel latch is in tight contact with the side of the notch in the count-wheel it will not release properly. In fact the count-wheel notch only lets the latch drop so that the hoop-wheel latch can drop into the gap in the hoop wheel to properly lock the strike cycle. The remedy is often just a matter of removing the ovality in the bushing holes.

Worn bushes aren't the only cause of striking problems but excessive bushing wear is the first

A rack-strike clock ready to commence the strike cycle for three o'clock; the reverse minute wheel pin has made contact with the lifting piece.

Most 30-hour clocks use a count-wheel mounted behind the back movement plate. The erratic striking in this centre-wheel clock was traced to excessive wear in the detent arbor (topmost of the three arbors on the left).

shows as commencement of the strike cycle a few minutes before the hour, with a prolonged repetition of strikes until the strike cycle proper commences.

Without the ability to pause the strike cycle on warning until the minute hand reaches the hour, the clock will simply proceed to strike the hour as soon as the lifting piece raises the

The badly worn pins on the pin wheel of this 30-hour are in need of immediate repair.

thing to look at. Below we have summarized just a few other faults, which could be put under the general heading of wear, loss and breakage.

Worn Pins on the Pin-Wheel

Wear is the inevitable consequence of the hammer trip rubbing over the pins of the pin-wheel at each hourly strike cycle (a clock strikes approximately 57,000 times every year). If the wear progresses uncorrected, the pin will eventually break, creating a space in the strike cycle.

Warning Wheel Pin Missing

Early count-wheel clocks of the Medieval and Renaissance periods did not use the warning arrangement that is common to longcase clocks. In those clocks that had only one hand or a fixed pointer against a revolving dial, an ingenious articulated lifting piece was used. In a longcase movement, the warning system prevents that premature commencement until the lifting piece has dropped. Consequently, the symptom of a detached warning wheel pin

A similar but later pin wheel; the pins have been replaced using stock pivot wire. The scribed circle is not just decorative, it is the maker's way of ensuring that all the pins will be on the same diameter.

hoop wheel detent. While the lifting piece is still raised, the count-wheel latch cannot drop, so the hoop wheel will not lock the strike train. When the lifting piece eventually drops off the pin, the train will lock on the next available count-wheel slot and the count-wheel will have lost synchronicity with the hour hand.

In a rack-striking clock, the strike cycle will commence a few minutes before the hour. Until the lifting piece has dropped off the reverse minute wheel pin, the rack hook will not engage with the rack, so the rack cannot begin to count off the strikes advance until the lifting piece eventually drops. The consequence is a delay of, typically, several minutes, during which the clock continues to strike unchecked.

A single pin on the warning wheel is struck every hour by the warning vane. They tend to drop out occasionally, especially after immersion in a cleaning machine.

A rack-strike clock on warning. The lifting piece/warning vane, which protrudes through the diagonal slot in the movement plate, has simultaneously released the rack and positioned the warning vane in the path of the warning wheel pin.

A birdcage movement with the striking train locked. The warning wheel is the topmost wheel of the strike train; the warning vane is below the bottom of the warning wheel rim.

The same clock on warning. The count-wheel detent is unlocked and the warning vane is raised into the path of the warning wheel pin to hold the strike cycle until the lifting piece has dropped.

The remedial work is quite straightforward. Hard steel pinion wire is available from horological suppliers.

Step 1 Examine and Prepare for a New Pin

First, establish whether the old pin has fallen out or just broken off, leaving the stump in the rim of the warning wheel. If necessary, punch out the broken stump from the back. The hole is cleaned up by broaching from the front of the wheel rim.

Step 2 Prepare and Fit New Pin

Use graduated twist drills to determine the required diameter of the new pin. The diameter of the appropriate stock wire will be slightly greater than the diameter of the hole. A short length is scored and snapped off the stock wire and a slight taper ground with a reasonably coarse stone.

Step 3 Fitting the Pin

Try the slightly tapered pin in the slightly tapered hole and, when satisfied that it can be pressed fully home, place the wheel onto a wood block and tap the pin firmly into place.

Lost or Broken Fly Spring

If the clock seems to strike very rapidly, it is quite likely that the fly is not correctly limiting the speed of rotation of the strike train. In other words, the vane part is free to spin loosely, relative to its arbor. In normal operation a strip of spring brass, soldered or riveted to the vane, exerts enough frictional pressure to allow the arbor to drive the vane. Occasionally the solder fails and the spring is lost.

Step 1 Examine, Measure and Prepare the Replacement

Check carefully how the original fly was intended to work and measure for the replacement part. It is more convenient to cut the replacement to the exact size. Brass shim plate is ideal for making up a replacement spring but, depending on its temper, it may require some light hammering to bring it to a sufficiently springy state. The diameter of the arbor is usually reduced over

All three flies have been repaired: on the left, a piece of clock spring has been riveted to the vane; at centre, a thin brass strip is effective; but on the right, the piece of soft brass wire roughly soldered on exerts no pressure against the arbor and was the cause of a rapid strike.

the width of the spring to locate the vane in one position along the vane.

The new spring should be filed to fit, so that when it is attached, the vane will be held in position on the arbor.

Step 2 Fit the New Spring

As a general rule, springs and heat don't mix, so riveting is probably a better method of fixing a spring to the fly. Soldering is best achieved with cleanliness. All traces of oxidation, dirt and grease should be removed from each surface prior to tinning. Do not use a blow torch because it will alter the spring temper of the brass. Instead, hold the spring tightly on the vane in its correct position with a pair of tweezers and use the tip of a soldering iron just enough to run the solder without altering the temper of the spring.

Loose or Broken Rack Tail

In eight-day clocks, the rack tail with its pin that falls onto the snail controls the throw of the rack and hence the number of strikes. It is designed to be adjustable on its mounting post. Just before every hour, the lifting piece unlocks

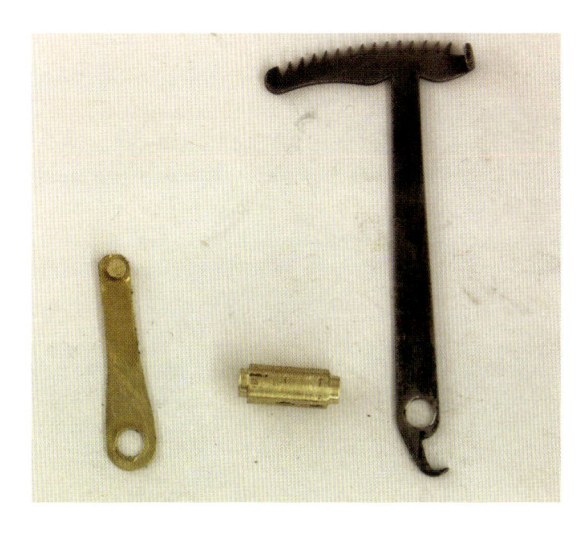

Tail, pipe and rack. The three components are joined by riveting the pipe to the rack and the tail. The riveted joint should be tight but with just enough movement to allow for fine adjustment of the rack.

the rack and it flies back until the pin on the rack tail hits the snail. Consequently, with 8,760 impacts each year, there is a tendency for the rack tail to shift relative to the post. The fault is often associated with a replacement rack spring, which, if it is stronger than the original, will make a harder impact.

Step 1 Tighten the Rack Tail
Place the rack on a hard surface and gently tap the rivet at the top of the mounting post to squeeze it tighter.

The result should be a fixture onto the rack pipe that will just allow movement, so that the fine tuning can be achieved.

Step 2 Fine Adjustment
Attach the hands with the dial removed and position the hand to approximately the five to twelve position. The clock starts the strike cycle by releasing the rack, which will fly back until the rack tail pin is stopped by the twelve o'clock step of the snail. With the lifting piece up in the warning position, the rack hook will also be raised but, as the lifting piece drops off the reverse minute wheel pin, the rack hook should drop squarely between two teeth of the rack. If not, the rack tail should be rotated slightly on its post. Then let the strike cycle continue by applying hand pressure to the drum and count off the strikes.

If all is satisfactory, turn the hands forward to the five to one position, the rack should fly back so the rack tail pin is stopped by the one o'clock step of the snail. As the minute hand reaches the twelve position, the rack hook should again fall squarely between two rack teeth.

The movement should then be set up on the test stand and allowed to run a few days.

Strike Cycle Fails to Commence

A common fault following re-assembly of a clock movement is caused by arranging the hammer trip to come to rest against one of the pins of the pin wheel at the end of the strike cycle. The effect of that misalignment of the wheels is to leave the hammer partly tripped at the end of the strike cycle.

In a rack-strike movement, the cause of the fault could be as simple as a wrong orientation of the gathering pallet during re-assembly. The remedy may be as simple as removing the gathering pallet and rotating; otherwise the

At the end of the strike cycle, the hammer should come to rest in a 'relaxed' position with no weight on the hammer spring.

The same clock with the gathering pallet incorrectly rotated by quarter of a turn, the strike cycle has been stopped by the rack pin against the gathering pallet but the hammer is stressed against its spring. At the next hour, the driving weight may not be sufficient to properly overcome the tension in the hammer spring at commencement of the strike cycle.

movement will require partial dismantling to correct the wheel meshing.

In count-wheel movements the hoop-wheel detent should drop into the hoop to lock the train just after the hammer trips, that is, just before the next pin of the pin wheel makes contact with the hammer trip.

WORN PIVOTS

Pivots and bushes wear excessively when they are not properly lubricated. In normal operation, a thin film of oil allows the pivot to float slightly in the bush but, as the oil dries, a rubbing line contact exists between the pivot and the bush, concentrating wear in both.

When they are examined with a magnifying glass, worn pivots have rough surfaces, often with ridges and grooves.

If the surface is only rough, it may be burnished with a steel burnisher, but pronounced ridges and grooves should be removed by first rubbing back flat with a fine diamond file, followed by a fine emery board. Attempting to burnish a very uneven pivot will result in a flaky surface on the pivot, which may break up, damaging it further and possibly causing harm to the bush.

A homemade Jacot drum is mounted in the lathe tail-stock. The groove in which the pivot rests during burnishing is exactly on the axis of rotation.

Burnishing by hand. A saw-cut in the wooden block supports the pivot, while the arbor is twisted with the fingers of one hand, and rubbed against the direction of rotation with the burnisher.

Step 1 Remove the Uneven Surface

In a lathe, the Jacot drum is used to support the pivot while it is abraded back with an emery or corundum buff and then burnished. Care should be taken to remove the absolute minimum of material from the pivot.

If a lathe is not available, the pivot is supported in a groove in a wooden block held in the bench vice. For a right-handed person, the arbor is supported between finger and thumb of the left hand and rotated against the direction of the emery buff pushed over the pivot with the right hand. A few strokes of the buff are usually sufficient to remove any unevenness.

Step 2 Burnish the Pivot

Whether burnishing in the lathe with the Jacot drum or on a wooden block, it is essential that the burnisher forces a cleanly defined and true shoulder at the root of the pivot. To achieve that geometric accuracy, the burnisher must be sharp and must be run exactly at right angles to the axis of the arbor. A few strokes of the burnisher should be sufficient to create a bright surface to the pivot.

Bushing wire and individual bushes are available from horological suppliers; bushes can be made up from any brass rod or bar.

WORN BUSHES

Wear in the bushes upsets the proper meshing of gear teeth and also increases friction and wear at the bearing surface that supports the pivot. The two effects combine to greatly increase the internal friction in the movement. Once bushes have started to wear, that wear spreads at an accelerating rate throughout the movement until either internal friction or some catastrophic failure causes it to stop.

Replacement brass bushing wire is available from horological suppliers in a range of diameters and length. Alternatively, new bushes can be made up as required from brass stock.

The looseness or shake of the pivots in their respective bushes should only be assessed after the pivots have been burnished.

Estimating when wear in bushes has become excessive takes a little practice and longcase clocks quite often function well with slack that might surprise model engineers. The easy preliminary way to test shake is to place the

The odd angles that these two arbors are lying at in their bushes suggests excessive wear. An attempt has been made in the past to close the bush on the left by punching a ring of holes. A legacy of previous mends of various qualities is a feature of most old clocks.

movement plate flat on the bench and stand each arbor in turn in its bush. If an arbor falls significantly off vertical, the bushing is obviously suspect. It may be found that the arbor falls off more in one preferred direction, suggesting that the bushing has developed wear in one direction.

Actually, wear in bushes is rarely distributed symmetrically and examination with a magnifying glass will usually reveal some degree of ovality in the bush, caused by the tendency for the meshing wheel and pinion to impose lateral loads that push the arbors apart.

Correction requires the insertion of a short length of brass bushing wire into a precisely formed hole in the plate and, as a matter of careful restoration of a historical artefact, the replacement bush should be of the least safe diameter. In other words, large enough that it will not rub through with normal wear but no larger than absolutely necessary. The engineer's term, interference fit means that the bush needs to be tapped or pressed quite firmly into place, but because the bushing wire is supplied with an axial pilot hole that will be opened out to form the new bush, it is essential that it is located for correct tooth meshing of the wheel and pinion. Consequently, before starting to open out the bush to take the new bushing wire, it is prudent to check the precise location of where the bush should be by using a depthing tool.

Bushes are fitted using simple hand tools or, alternatively, a specially made bushing tool, normally used by clock-menders, speeds the process and gives a consistent and reliable result.

Hand-Bushing

Step 1 Check the Proper Location of the Bush

Set up the arbor to be bushed with the driving arbor in the depthing tool and, using the adjusting screw, gently close the distance between the arbors until meshing begins to feel uneven or rough. Wind the adjustment back out, feeling the quality of the meshing by continually rotating the arbors. Once the proper depthing has been found (i.e. the greatest depthing or closest centre-to-centre distance with free meshing),

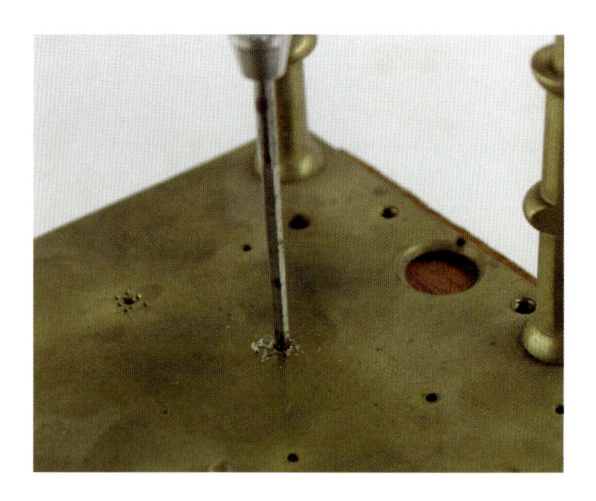

Broaching for a new bush. It is important to keep the broach perpendicular to the movement plate and to ensure that the centre will be correct for meshing.

the centre-to-centre distance is transferred to the movement plate as a scribed arc.

Step 2 Remove Ovality in the Worn Bush

A wear lobe develops as the pinion is forced against the side of the bushing. Using a round needle file, remove a lobe from the side of the bush exactly opposite to the oval wear. The effect will be to exaggerate the ovality but, once the hole is prepared in this way, the geometric centre of the oval hole will coincide with the arc previously scribed by the depthing tool.

Step 3 Measure the Pinion and Decide on a Diameter for the Replacement Bushing

As a conservation matter, the bushings should be planned so that the minimum original metal is removed and, when future re-bushings are required, only the replacement bush will be replaced. The ideal diameter of the bushing wire is approximately twice the pivot diameter.

Step 4 Broach the Hole to Take the Bushing Wire

The hole in the movement plate should be broached from the inner face because broaching gives a slightly tapered hole, so in the event that the bush comes loose, it should tend to slide into the movement, not out of it.

The diameter of the pivot was 1.710mm, so a bushing wire of 3.50mm diameter (approximately twice the pivot diameter) was chosen for the replacement.

Care should be taken to keep the broach perpendicular to the movement plate

There are several ways of getting the broached diameter correct and the first is by simple trial and error, trying the bushing wire as the diameter is gradually increased until it can be tapped into the hole. The second method is by measurement. If the usual five-sided broach is used, the hole diameter formed is greater than the distance from an edge to the opposite flat. The factor is 1.106, so with the micrometer set to the diameter of the bushing wire, less about two-thousands of an inch, divide by 1.106, set the micrometer to that dimension, slide the broach into the micrometer until it just fits and then mark that position with a spot of Tipp-Ex or similar.

During the broaching operation, the hole will enlarge as the tapered broach cuts deeper and when the spot reaches the movement plate, the hole will be the desired diameter.

A slight complication occurs if the movement has been re-bushed in the past. It is quite likely that the previous bush will drop out as the hole is broached, in which case the diameter of the intended bushing wire should be checked. To conserve the movement as far as possible, it is preferable to replace the bushing and avoid removing any further original metal.

Step 5 Fix the New Bush

First, remove any burs from the end of the bushing wire, and then cut the new bush. If it is judged necessary to rivet the bush, an allowance is made for the extra length needed for riveting.

The bush is tapped smartly into place. If it is not tight, it just won't do and, in that case, the hole should be re-broached for the next larger bushing diameter.

With the inner face of the bush flush with the movement plate, it is stoned flat, taking care not to remove metal from the plate.

Turn the plate over and file away any excess length and stone flat and buff to finish.

The new bush is cut slightly longer than the thickness of the plate.

The new bush is driven into the plate from its inside face, while the plate is supported on a brass anvil.

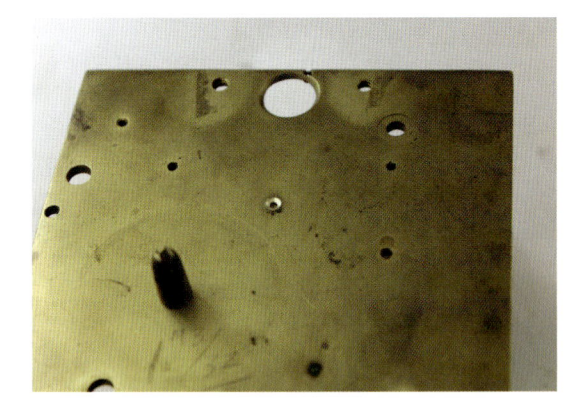

The bush is rubbed back to be flush on the inside of the movement plate and a hemispherical oil sink is cut on the outside.

Step 6 Cut the New Bush Diameter to Fit the Burnished Pivot

The pilot hole in the bushing wire is opened up with a series of broaches until it takes the pivot. Initially, the cutting is done from the inside and care should be taken not to cut too much, by trying the pivot frequently until an easy fit is achieved.

Once the pivot fits with a small degree of shake, the hole is broached very lightly from the outside and then burnished with a smooth broach. Finally, the oil-sink is cut on the outside.

Step 7 Test the Work

When the bushing is complete, the arbors of that part of the movement are re-assembled and the plates are pinned together, so that the wheel-work can be checked with light finger pressure to make sure that all runs freely.

Step 8 Clean Ready for Re-Assembly

It is essential that all traces of metal swarf are removed by a very thorough cleaning before the movement is properly re-assembled.

Using a Bushing Machine

The Bergeon or similar bushing tool is used by professional clock-menders because it speeds up the re-bushing process; it is used in conjunction with Bergeon bushes or a Bergeon bushing rod.

Step 1 Set Up the Movement Plate in the Bushing Tool

When a pivot hole has been chosen to bush and, if necessary, the corresponding pivot has been filed and burnished, the hole is centred in the bushing tool by placing the movement plate between the two clamps, with its inner face upwards and lowering the centre into the hole. The plate is moved until the hole is dead centre and the clamps tightened.

Step 2 Reaming the Hole to Accept the Bushing

A bushing with the smallest possible outside diameter, and of suitable height for the thickness of the plates, is selected. Check that the pilot hole is slightly smaller than the pivot. The outside diameter of the bushing corresponds to the size of a standard reamer supplied with the bushing tool. That reamer is to be used last, after the hole has been opened in a series of steps using reamers of increasing diameter.

The reaming is achieved by first placing the reamer in the mandrel and tightening with the grub screw. Turning the handle of the mandrel gradually with light pressure in a clockwise direction, as the reamer enters the hole, results in an exactly circular hole size of the correct diameter, at the correct location in the movement plate.

Step 3 Fitting the New Bushing

The plate is removed from the bushing tool and the stake that was used to support the plate during reaming is replaced with a plain stake. The hole is re-centred and a driving punch is fitted into the mandrel. (The driving punch should be slightly larger than the bushing.)

The bushing, which is supplied by Bergeon, is placed square over the hole ready to be tapped into place. The mandrel is made to be tapped with a medium hammer to drive home the bushing. If the bushing is too long and protrudes above the inner face of the movement plate, it can be cut back using a pivot cutter of suitable size, being careful not to damage the plates.

Step 4 Broaching the Hole to Accept the Pivot

The Bergeon bushing tool is not used for

adjusting the diameter of the hole to the pivot size – that must be done by using a series of tapered broaches (as per step 4 of the hand-broaching procedure).

Although the bushing tool is designed to speed the process, it is not an absolute requirement and, in fact, there are instances where it cannot be used; for example, the bushing in the back-cock that supports the back end of the pallet arbor must be replaced by the hand process because the bushing tool cannot support small pieces, such as a back-cock, without adaptation.

The novice, attempting pivot and bushing restorations for the first time, would be well-advised to gain a little experience by practising on cheap, modern movements before working on valuable heritage clocks. However, once the technique is mastered, he or she can look forward to the satisfaction of achieving those sorts of results that used to be the sole prerogative of professional clock-menders.

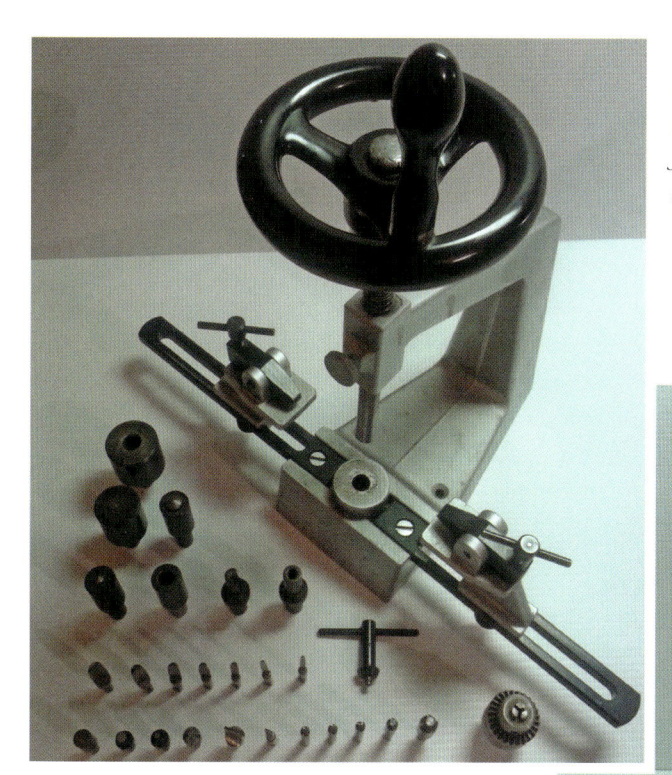

A Bergeon bushing tool with its accessories. The movement plate is clamped to the frame and the arrangement guarantees that the bush will be perpendicular.

With the movement plate lined up it is just a matter of positioning the new bushing and driving it into the plate with the mandrill.

Chapter 6

More Difficult Movement Procedures

The procedures described in this chapter are difficult but not impossible for the novice, nor are they particularly demanding in terms of workshop equipment. Nevertheless, like all work involving clock movements, they require care and patience, and quite a lot of practice.

CURING RATCHET AND CLICK PROBLEMS

General Arrangements

Engineers call them ratchets and clockmakers call them clicks – they are the simple one-direction devices comprising a sprung pawl that allows the drum or sprocket to rotate relative to the great wheel, while the movement is being wound. Consequently the driving weight is supported only by the click, and any failure is apt to result in a catastrophic uncontrolled drop of the driving weight with the likelihood of damage to the movement. As with other aspects of longcase clock movements, clicks vary in design detail and it is prudent to properly understand the specific part prior to attempting a repair.

Going train; no click – driving force continuous

Strike train, the single click allows re-winding

Counterpoise or counter weight keeps the rope taught on the great wheel sprockets

Pulley supports the single driving weight

LEFT: *Huygens is credited with the invention of the endless rope system in 1658. It removes the click from the going train.*

The early steel circular spring on the left is generally more robust than the later type, which uses a pin-jointed pawl and a strip spring.

More Difficult Movement Procedures 115

The fundamental design flaw in eight-day clicks places all the weight of the driving force on the pin that fixes the pawl to the wheel rim.

It is not widely realized that when Christian Huygens invented the endless rope system for driving two trains of wheels from a single weight, he had made possible the simple maintaining-power system.

In a typical endless chain or rope 30-hour clock, there is just one click, which is located on the front face of the strike-side sprocket. Although they invariably act on the spokes of the great wheel, there are two alternative designs: the early type is a circular spring with a step, which acts as the pawl; and the later type, a separately sprung pawl, which is inherently more prone to mechanical failure.

An eight-day movement is arranged a little differently because the driving cord spools onto a drum; a ratchet gear ring is formed around the edge of one of the cheek plates of the drum and the pawl is mounted on the face of the great wheel, forced by a spring into the ratchet teeth.

In order for it to work as intended, the click should be securely mounted but free to move. In operation, the spring should snap the pawl back smartly as it drops off each tooth or spoke. The most common problems with clicks are:

- Excessive movement and, occasionally, complete detachment of the pawl arising from wear in the mounting hole in the great wheel (applies more to eight-day clocks).
- Weak or broken springs (which may be of brass or steel).
- Damaged teeth on the ratchet gear (applies more to eight-day clocks).
- Wear on the crossings or spokes of the great wheel (applies to 30-hour clocks).

Loose Pawl

Step 1 Identify How the Pawl is Mounted

Various methods are used to mount the pawl on the great wheel of an eight-day movement. If the stub is solid with the pawl, formed from one piece of metal, it is often riveted into a countersink on the rear face of the great wheel, but it may be threaded and screwed to the great wheel. The latter method of attachment is the one most prone to excessive wear during normal operation. The best procedure for removing the pawl will become apparent once the type of fixing has been established using a magnifying glass.

Step 2 Remove Old Pawl and Inspect Mounting Hole

For screwed pawls, it is usually necessary to first remove the spring to allow full rotation

The riveted end of the click-retaining pin must be removed to extract the pin for restoration work.

Another intrinsically flawed design is the threaded pin type of pawl. A screw thread is a useful fixing device but in this application, where the threaded parts move during normal operation, wear is inevitable.

of the pawl. With the spring out of the way, and the pawl unscrewed, the nature of the wear may be examined. Excessive wear to the internal thread in the brass great wheel may make it necessary to relocate the pawl in a new hole. Otherwise, if there is enough metal, the hole may be broached out to take an oversize stub, either threaded or plain riveted.

When a riveted pawl is loose, the source of the problem is often slackness from wear at the seat of the rivet, which causes the rivet itself to deform as the pawl stub rocks in its hole. Old riveted stubs should be centre-punched and drilled to remove the rivet flange.

Again, the condition of the hole and its circularity should be inspected and judgements made about whether or not to prepare a fresh hole (which will naturally mean re-locating the spring).

Step 3 Make and Fit a Replacement Pawl
Replacement pawl blanks may be bought from horological suppliers; they are formed from approximately 3mm mild steel sheet and, generally, the stub is threaded. Like all stamped steel blanks supplied by horological suppliers, they are only blanks, intended to be shaped exactly for the individual clock movement.

Alternatively, in the workshop, a new pawl blank may be formed from a piece of 3mm mild steel plate cut roughly to a T-shape. After heating and bending the blank to shape, it is mounted on a wax chuck to turn the stub to the correct diameter. Alternatively, a piece of 2.5–3mm thick steel or brass is cut to shape and drilled to accept the stub, which is turned with a shoulder and riveted into a countersink.

As the pawl is riveted onto the great wheel, there is a risk of the rivet being squeezed too tightly and locking the pawl onto the great wheel. To avoid that situation, a clearance is provided by temporarily placing a couple of pieces of old suspension feather (or any 0.2mm shim) between the pawl and the wheel prior to riveting. Once the stub is riveted, the feathers are removed and the pawl should rotate freely.

Springs

Click springs are formed from heat-tempered, high-carbon steel or work-hardened brass. Blanks are obtainable in an annealed or non-springy state from horological suppliers, or may be formed in the workshop. The procedures for work-hardening brass to make it springy and for heat-treating high-carbon steel are described

The click pictured on page 115 was quite weak and on dismantling was found to exert hardly any pressure on the pawl.

The spring was removed from the wheel by slipping a sharp blade under and prying it up. After some hammering to bring to a better spring temper and even out the curve, it was offered up to the wheel (notice the two fixing lugs on the left) and found to be too long because the free end acts too close to the pin.

With the temper and length of the spring corrected, it was re-fitted to the wheel with a spot of thread locking fluid on each of the retaining lugs.

in Chapter 4. In the interests of the historical integrity of the clock, it is appropriate, and possibly even important, when making replacement springs to use the original as a pattern, both in the metal used and the shape.

When it is completed, the spring is fixed to the great wheel; so, if it has been made from steel, any fixing holes should be drilled prior to heat treatment.

The great wheel is re-assembled and the click tested before re-fitting to the movement.

MAKING REPLACEMENT HANDS

The style of hands evolved over the longcase-clock period so, on the one hand they provide a useful dating but on the other hand, replacement hands of the wrong style can easily create an irksome blemish. The importance of a good set of hands cannot be over-emphasized and, consequently, we have included here some fairly detailed guidance about making up new hands.

As a conservation matter, if the original hands are lost, it is appropriate for the conserver to provide replacements that are close enough to give a sense of the original, without creating a false impression.

Hands are fragile and susceptible to serious damage or loss and occasionally evidence of previous restoration work is found in inappropriate substitutes. Making new hands of the correct style and material is yet another everyday part of clock restoration. Given the necessity to fit new hands, the first requirement is to decide upon a style that will suit the clock dial. Replacement hands of varying quality may be bought from specialist makers or horological suppliers. At the one extreme of cost, when carefully made by a skilled craftsman, they can be as good as the originals, but at the other extreme, poor imitation, stamped hands of an unsympathetic style are often just disappointing blemishes on a good dial.

In early longcase clocks with brass dials, the minute and hour hands should be of different designs. Matching hands, originally of steel, only became fashionable relatively late in the development of clocks, at about the same time as painted dials were introduced and the use of steel was subsequently supplanted by brass.

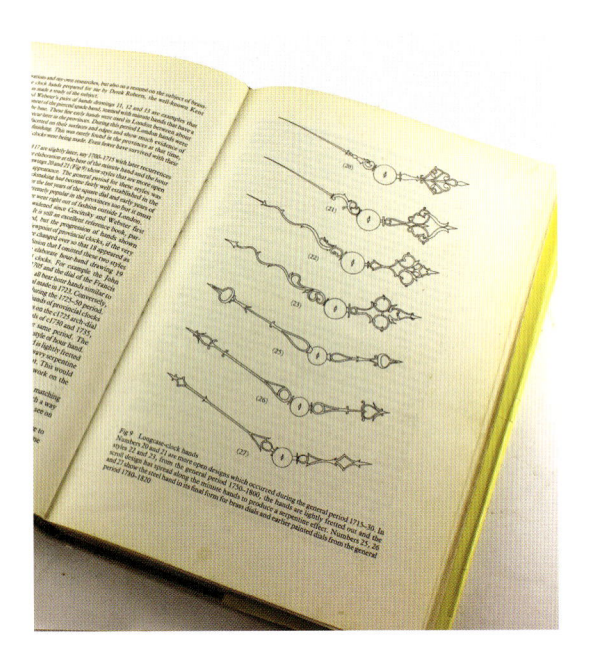

More Difficult Movement Procedures

The evolution of hand styles has been well-documented, originally in 1913 by Cescinsky and Webster and later by Brian Loomes, Richard C. R. Barder and Darken and Hooper, etc., and reference books should be used carefully to design replacement hands that are in sympathy with the dial.

As an alternative to copying pictures of hands, it is possible to use the hands of another clock as a pattern, as long as the period and styles are the same and the dimensions of the chapter ring are identical. The tip of the hour hand should just reach into the bottom of the hour numerals and the minute hand should just cover the minute marks.

Dials and chapter rings vary in size; having found the correct design, it may be necessary to scale an image and print and paste onto the blank.

Steel Hands

Step 1 Choose and Prepare a Suitable Steel Blank

Having decided on the style for a replacement hand, the next consideration is the material, thickness and length.

For clocks of the early painted-dial period, almost any piece of good-quality steel, about 1.5mm thick, can be used; but a sample should be tried first to make sure that it will cut, saw and file and polish, and will blue with heat satisfactorily. (Mild steel is a more recent innovation and mostly unsatisfactory for longcase clock hands.)

Clock hands are well-documented and a number of books are available that show how styles evolved over time. Richard C. Barder's English Country Grandfather Clocks *is just one such book.*

The early eighteenth-century clock pictured here had been fitted with poor-quality pressed steel hands. A pattern for a suitable replacement set was found on another contemporary clock. Notice the position of the tips of the hands relative to the hour and minute markings on the chapter ring.

A paper pattern is pasted onto a high-carbon steel blank, which preferably thins towards the tip.

For earlier clock hands there is another consideration. By looking carefully at early hands it will be noticed that they tend to be much thicker at the mounting boss than the tip. Typically, the thickness reduces from up to 2.5mm at the boss to 0.5mm at the tip.

A suitable blank can be made up by forging a strip of high-carbon steel. Reddening and hammering will reduce the thickness, but as an alternative it is relatively easy to find steel tools that have the property of thinning to about 30 per cent over distances of 10–20cm. Worn shovels and mason's trowels are examples.

If the blank is not already annealed, it is heated to a dull cherry red and allowed to cool slowly, which will render it soft enough to be cut and filed.

Step 2 Mark Out and Rough Out the Hand

If the design is to be traced from another hand, the blank is first painted in white undercoat paint and, once dry, the outline is marked with a fine mapping pen. Alternatively, the design is printed onto paper, which is glued onto the steel blank.

All of the piercings are drilled and the sheet is placed into a vice or finger plate ready for cutting out. Using a fine, flat blade in a piercing saw, the hand is cut out roughly. It is a matter of personal preference whether the piercings are cut first or after the outline of the hand has been roughed out.

Step 3 Filing and Polishing

Needle files are used to smooth and shape the outline of the hand. In order to prevent

A junior hacksaw will cut the rough shape quite quickly. Good support in a vice or finger plate is essential.

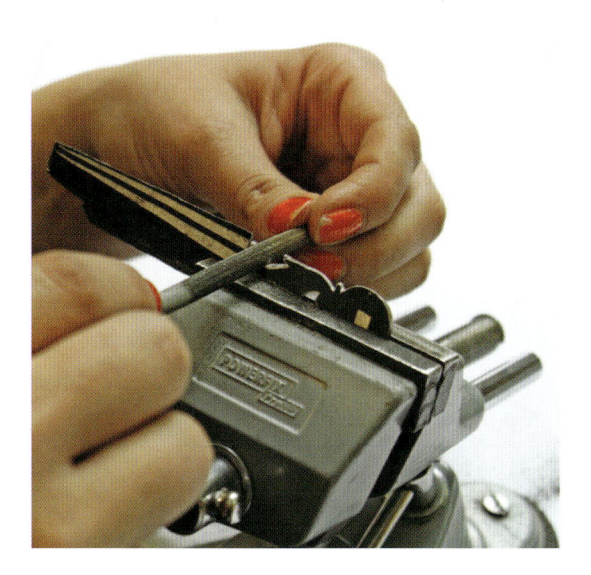

Completing the detail nearest the boss. It is usually easier to work from boss to tip.

Steel files complete the shaping followed by finer diamond-coated files. The marks are removed by buffing with fine corundum paper before polishing with chrome polish.

Blueing in a flame takes practice and patience but the results are well worthwhile. Work from the large areas towards the points because the heat tends to spread from the point of application.

accidental damage during that process, the part of the hand being worked is kept as close to the chops of the vice as possible. A finger plate is useful to support the hand during filing.

Since the hand will be blued, it is necessary to polish the flat surface to a mirror finish by gradually taking it down with grades of corundum or emery paper and finishing with a fine, hard metal polish, such as chrome polish.

Step 4 Blueing

Steel hands are not engraved; the historically correct treatment is blueing by heating the polished hand in the same manner as tempering, to achieve a slightly purplish iridescent dark blue finish.

There are two approaches to blueing:

- The hand is heated slowly in the flame of a spirit lamp and, as the colours develop from yellow through brown to purple, the position of the hand in the flame is adjusted to keep the colours even. As the brown colour changes to purple/blue, the hand is removed from the flame and cooled in blowing air to stop the colour continuing to turn to an opaque blue/grey. It takes practice, but in the event of a disappointment the hand can simply be repolished and the blueing repeated.

- A blueing tray is used. The hand is placed in a shallow steel tray on a bed of about 1cm of brass swarf and heated above a gas flame. The colours will appear on the polished surface and the hand should be removed quickly and cooled in blowing air or wiped with an oily rag before the blueing goes too far.

Brass Hands

Step 1 Mark Out, Prepare and Rough Out the Hand

Brass longcase clock hands (the norm for clocks dating from about 1820) are always in matched pairs, so if only one hand is missing, the remaining hand should be used as the pattern for the replacement.

The material to use is 1.5mm-thick ordinary or common brass sheet, but before attempting to cut out the rough shape of the hand, the blank should be tempered by hammering lightly to bring it to a hard or extra hard temper.

The piercings are drilled first and then sawn with a fine piercing saw (for brass, a very sharp blade is preferable).

Finally, the outline is roughed out using a fine hacksaw, taking care to position the blank on a bird-mouth to protect it from bending stresses.

Squared paper is useful for making up patterns. The length from boss to tip is measured from the dial and the overall design is taken from the remaining hand (brass longcase clock hands are always in pairs). Once marked, the pattern is pasted onto the brass blank.

A bird-mouth is ideal for cutting around the detail.

Step 2 Filing Out and Engraving
The rough cut outline is filed back to the final profile and, once that has been achieved, the sharp edges are filed back and the face of the hand is engraved to break up the otherwise flat appearance.

Step 3 Apply a Lacquer
Lacquer, which is just shellac or French polish, is used to prevent the surface of brass from tarnishing. It is applied to a brightly polished brass surface by heating the brass to hand hot (about 60°C) and painting the diluted shellac polish on with a soft brush. Only a very thin coating of shellac is required to protect the surface; the brush should be just damp with polish, not dripping.

Gilding was used on brass hands and gold leaf can be applied onto clock hands using ordinary gold size and either pure gold or synthetic (alloy) leaf. If alloy leaf is used, it should be sealed with shellac immediately to prevent tarnishing

REPLACING A BADLY WORN PIVOT

When a clock has been allowed to run for a long time without oil, the pivots naturally become worn unevenly and occasionally the wear is so great that the pivot is no longer serviceable; in the extreme case pivots completely snap off. Fitting a new pivot should not be considered as sound craftsmanship but, if the arbor, and especially the pinion, are in good condition, pivot replacement may be the best option for both expediency and historical integrity.

If the replacement pivot is to be located next to a pinion, it is more important to arrange it exactly on the axis of the pinion. In other words, the pinion must not become eccentric.

Without Annealing

Drilling a hardened steel arbor is difficult or impossible with steel twist drills, and using very hard tungsten carbide bits should only be used as a last resort. Consequently, the usual procedure is to first anneal the arbor and then re-harden after drilling. Sometimes annealing is just not a practical option, so the drilling has to be done very carefully, setting up the work so that it rotates with no shake. Centring is critical and a very sharp graver is used to slightly indent the squared end of the arbor, so that drilling will be exactly on the axis of rotation.

Once the arbor is drilled, the pivot is made up from stock pivot wire, which, when it is stoned down to the hole diameter, is left with a roughened surface.

To avoid the heavy look of large areas of brass, the surface is broken up by engraved details, which can be punched or stamped quite easily.

This pinion has worn far enough to require replacement. The clock dates from the late seventeenth century and the one-piece wheel and collet was silver soldered to the arbor, so removal of the wheel could not be justified on conservation grounds.

The arbor was set up in the lathe with a temporary steady to support the free end. The steady comprising a split wood block, drilled to the diameter of the arbor, is held firm in the tool post. The 1.2mm tungsten carbide bit is run in about 5mm.

Once fitted, the new pivot is likely to serve a further three hundred years with the right maintenance, and the minimum intervention principle of good conservation practice has been followed.

The square end of this pallet arbor has broken off and it has been decided to drill the arbor and insert a new squared end.

The arbor is measured and drawn in AutoCAD with the new piece designed in. The time taken in preparing a simple drawing is made up during machining.

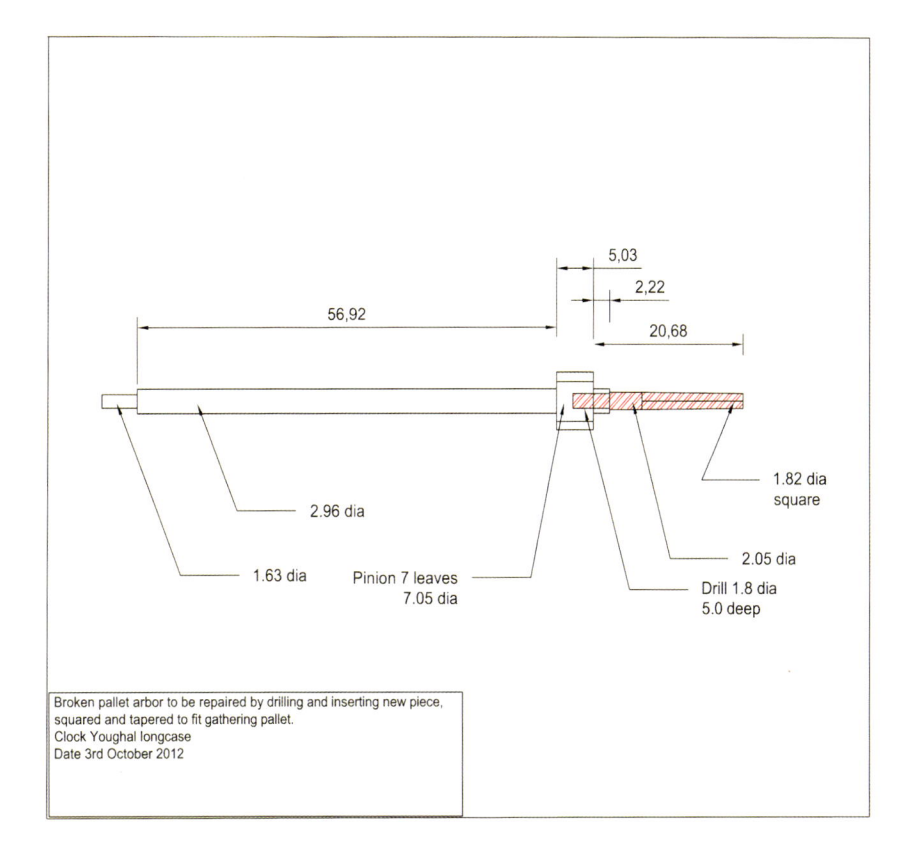

```
5,03
2,22
56,92
20,68

2.96 dia
1.63 dia    Pinion 7 leaves
            7.05 dia
                           1.82 dia
                           square
                           2.05 dia
                           Drill 1.8 dia
                           5.0 deep

Broken pallet arbor to be repaired by drilling and inserting new piece,
squared and tapered to fit gathering pallet.
Clock Youghal longcase
Date 3rd October 2012
```

The replacement pivot is tapped gently into position with a drop of thread-locking fluid. (If the fit is too tight, there will be a risk of bursting the end of the arbor as the pivot is driven in.)

Step 1 Measure and Record the Part and Plan the Restoration Work

Any old clock has a legacy of repairs, and work done today will become part of tomorrow's legacy. In the interests of careful conservation, especially on rare and important clocks, any work should be recorded. Drawings will also be helpful in designing the work.

Step 2 Prepare the Arbor and Remove the Old Pivot

Prior to attempting to drill the arbor, it must be softened sufficiently by heating to near red. It is then placed into the lathe, the old pivot is removed and the cut end stoned flat.

Step 3 Mark and Drill the Arbor

The traditional method is to find the centre of the rotating arbor and mark it with the point of a graver, just deep enough to start the drill hole accurately. Alternatively, a centre bit is fed lightly into the end of the arbor, enough to make a conical depression.

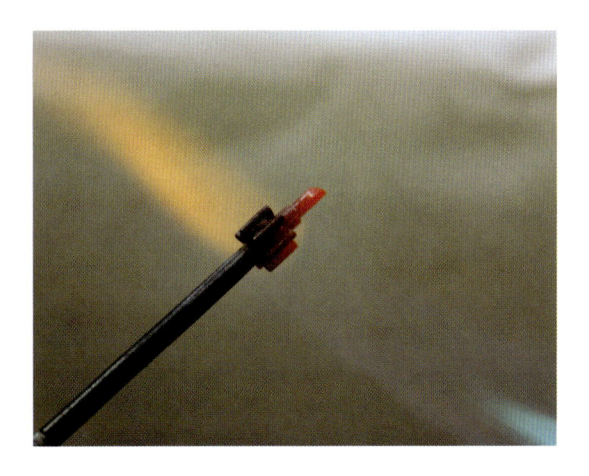

Once the part has reached red heat, it is allowed to cool slowly.

A centre bit is made to be rigid enough so that it will not wander over a flat surface.

The twist drill should be well-sharpened before attempting to drill the hole. The easiest way with small drills is to support the drill in a pin vice and rub back on a fine diamond abrasive block. Blunt drills tend to snag and snap off in the hole!

Step 4 Drill the Arbor and Fit the New Pivot

First, decide on an appropriate diameter for a new pivot. Although some clocks use a uniform tooth size, in most, both tooth and pivot sizes decrease from the great wheel to the escapement. Other than on the great wheel arbors, typical pivot diameters for longcase movements are between 1 and 2mm. The diameter of the replacement pivot should be reasonably close to that the original.

Pivot wire is obtainable in a range of sizes from horological suppliers. Otherwise, turn down a length of silver steel or even a piece of an old arbor.

To obtain an interference fit with a limited range of drill sizes, drill the hole to the size below the pivot wire and then reduce the pivot wire in the lathe with a fine file, forming a slight taper so that it fits about half way into the drill-hole in the arbor. Once the metal shaping has been finished, the pivot is fitted to the arbor. Traditionally, the new pivot was thinly coated with a very fine abrasive dust, but an effective modern alternative is Loctite or a similar thread-locking compound. In each case, the new pivot is tapped fully into the hole.

Step 5 Check for True

The arbor is set up between lathe centres and turned slowly to check that it runs true and even. If a wobble is detected in the arbor, the high side is tapped on a stake with a cross peen hammer. It is useful to use chalk held on the tool rest to detect eccentricity in the arbor; the chalk will leave a mark as it contacts the high part.

Step 6 Re-Temper the Complete Pinion

Finally, the parts are wrapped in iron-binding wire and then the whole is heated to a dull cherry red and quenched. The binding wire is removed, the arbor is cleaned and buffed, and then tempered back down to dark-yellow turning to red (about 250°C). Care should be taken not to concentrate the heat on the pivot, which will soften it unduly.

RE-FACING ESCAPEMENT PALLETS

As each of the pallets of the escapement rubs across a tooth of the escape wheel nearly sixteen million times every year, inevitably the rubbing action will eventually cause a characteristic wear track to develop on the face of each pallet.

When the new part is fitted into the arbor, the assembly should be quench-hardened and then polished, so that it can be tempered up to a brown/red colour.

The repair options are:

- Move the escapement — sometimes the pallets are wide enough to allow the escapement to be moved slightly along the arbor to present a fresh rubbing surface, but usually, re-surfacing is a safer option than stoning the pallet surfaces back to remove traces of wear tracks.
- Stone back the pallet surfaces to remove the irregular wear surface. However, removing metal to flatten the pallet faces may be a pointless cosmetic exercise because of the alteration in the distance between the pallet points and consequent change in depth of meshing in the escape wheel.
- Re-surfacing the worn faces of the pallets — the lost hard steel surface is replaced with a layer of similar material soldered in place.

Step 1 Thorough Inspection
It is quite likely that a worn escapement will be uneven in action. So, before any work is done, both parts of the escapement should be examined together.

The easy way is using a depthing tool, which has been set at the spacing taken from the movement plate. The usual spacing for the escapement pallets in a 30-tooth wheel is seven and a half teeth (i.e. a half tooth nearest to a quarter of the tooth count); it could be six and a half or eight and a half, or any number, but it's always a half because half a tooth is one tick.

Step 2 Prepare the Worn Pallets for Re-Surfacing
Remove the wear tracks by stoning back the faces, keeping the surface flat across the pallet and maintaining the vertical curvature. Using a small spirit lamp flame, the re-shaped pallet faces are tinned ready for soldering.

Step 3 Prepare the Replacement Wear Surface
From a piece of old steel clock spring of a suitable thickness (between 0.25mm and 0.5mm), a strip is cut the width of the pallet. Using two small pairs of pliers or a pair of turning pliers, the end of the strip is turned to the curvature of the pallet and, after rubbing with emery, it is tinned ready for soldering.

Step 4 Solder the Replacement Wear Surface to the Pallet
With the escapement anchor held in a vice, and the patch held in contact, the spirit lamp is held under the pallet, just until the solder runs. Excessive heat at this stage must be avoided, as it will anneal the new wear surface. After

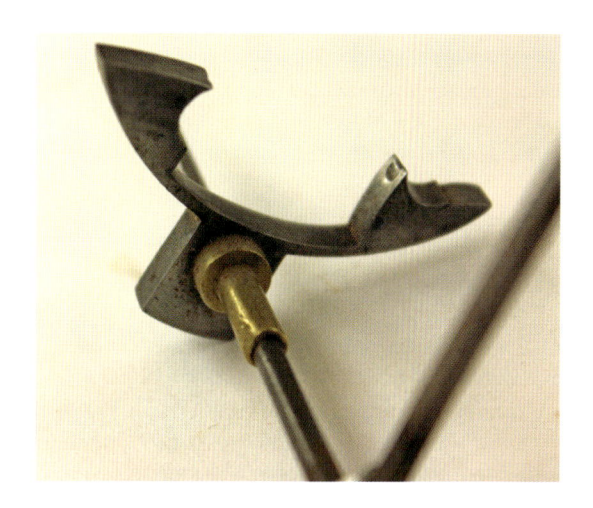

Once wear has become noticeable, it is likely that the efficiency of the escapement will be compromised.

After the worn faces are stoned flat, they are tinned. Strips of clock spring are cleaned and tinned and then soldered to the prepared faces.

The re-surfaced faces are first trimmed back with a coarse abrasive stone (upper pallet) and then (lower pallet) trimmed and finished with a diamond coated file.

cooling, the excess piece of spring is cut off with tin shears.

Step 5 Repeat the Re-Surfacing Process on the Second Pallet

Step 6 Trim the Excess and Finish

The rough, cut end of the spring is stoned back to the desired profile, which will be one that fits the escapement wheel for curvature and distance between pallet points. The escapement is repeatedly checked using the depthing tool between stoning until it appears correct. The pallets are then buffed with fine emery and polished to a bright finish.

REPLACING A BROKEN WHEEL TOOTH

Wheel and pinion teeth are always prone to breakage, especially when meshing is thrown by wear at the pivot bushings. A high proportion of breakages is attributable to casting imperfections in the brass, but any mishap is likely to cause tooth damage. The teeth most vulnerable are those at the bottom of the wheel train, where forces are greatest.

Although it is not feasible to replace a missing steel pinion leaf, brass wheels may be repaired by soft soldering a new piece of brass and shaping to suit.

Traditionally, the new piece was dovetailed into the wheel, with a sliding fit from the side, and soldered in place. Alternatively, a parallel slot is cut in the wheel and a corresponding tight-fitting piece pressed and soldered in place.

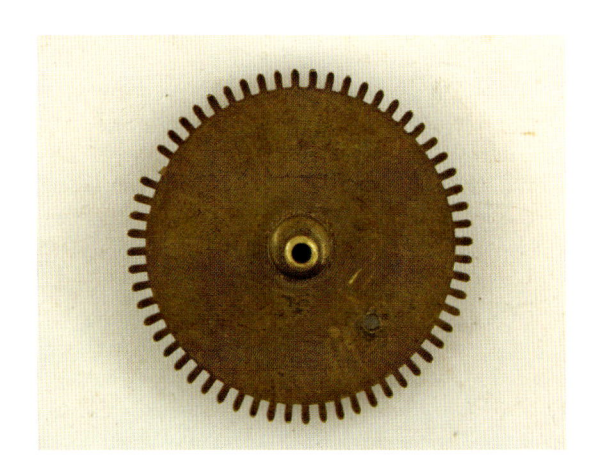

A missing tooth on this wheel, from the date work of an early painted dial clock, requires replacement.

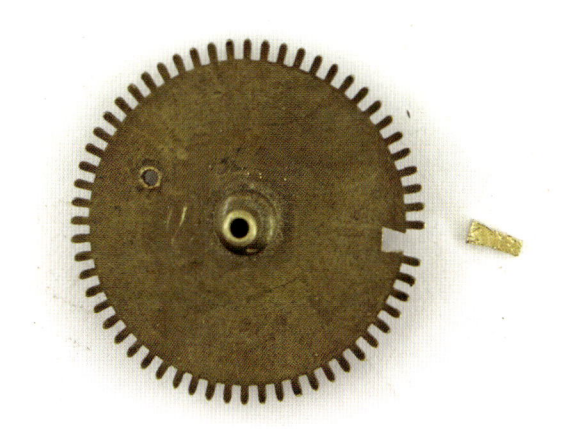

A dovetail-shaped slot is cut in the wheel to take a replacement tooth. The replacement is cut from a piece of hardened brass sheet.

The tinned tooth is tapped into place.

Step 1 Clean the Wheel
The effectiveness of soldered repairs relies on preparation, including the removal of all traces of dirt and oil from the vicinity of the repair.

Step 2 Prepare the New Tooth Blank
From a piece of compo brass sheet, reduce the thickness to that of the tooth shanks in the wheel. The blank should be work-hardened to hard or extra hard temper.

Step 3 Prepare the Wheel
The most common location for a tooth breakage is flush with the root. Otherwise, the remaining stump of the shank must be removed by filing down flush with the root.

A saw-cut is made, radial to the wheel at the mid-point of the gap, extending to at least one and a half times tooth thickness. The initial saw-cut is carefully opened to the thickness of the replacement, taking care to keep the opening exactly central and the sides square and true.

Step 4 Solder in Place, Trim and Finish
The mating surfaces are tinned with a very thin skim of soft solder, which may require rubbing back to achieve a tight fit. Once the new tooth is soldered in place, it is cut to height and the addendum filed to the same shape as the adjacent teeth.

The repair is buffed back to the surface of the wheel and the tooth profile finished by filing and buffing.

The result should be mechanically serviceable, so care should be taken to achieve the correct tooth profile.

Chapter 7

The Most Difficult Workshop Procedures

MAKING REPLACEMENT WHEELS

Sometimes brass-clock wheels get lost or become so badly damaged that replacement is the only option. The wheels inside the movement plates, especially those lower down the train near the great wheels, transmit more power and tend to sustain damage (with the loss of two or more consecutive teeth, replacement of the wheel is the preferred option). Wheels outside the movement plates, the motion-work and date-work, are more likely to be lost. Date and moon-phase wheels were often deliberately removed in order to improve a temperamental clock.

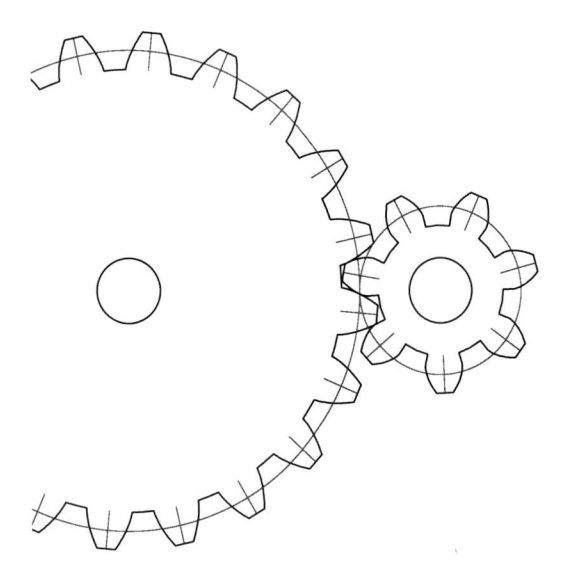

The involute tooth form developed with the motor industry: it conforms to modern ideas about how gear teeth should be designed. It is best suited for transmitting power from fast pinions to slower wheels, as in motor car transmissions.

The prospect of making up a new clock wheel may seem a little daunting to the novice but it is worth considering the facilities available to the makers of longcase clocks, let alone clockmakers in the thirteenth century, compared to what is easily available today. (Actually, an astronomical instrument containing many intermeshing gear wheels and dating from 100BCE was discovered in a shipwreck off the island of Antikythera in the year 1900.)

There are two aspects to wheel cutting: first, dividing the wheel into the correct number of equally spaced teeth; and, second, cutting the gaps uniformly between the teeth. The objective of properly designed gear wheels is that the meshing teeth roll over each other with a minimum of resistance. In essence, the thickness of the tooth should be slightly less than the width of the gap between teeth on the opposite wheel, and each pair of meshing teeth roll together with negligible friction.

SOME BACKGROUND: INVOLUTE OR CYCLOIDAL OR NEITHER?

Cutters for involute gears are quite easy to find – they are used extensively in small machinery and by model engineers. Involute gears have no role in longcase clocks and, in general, it is a mistake to attempt to make replacement wheels for longcase clocks using involute gear-cutters. Cycloidal gear teeth, on the other hand, tend to be similar to those found in longcase clocks. However, it should be clearly understood that longcase clock wheels mostly do not conform to any standard form and, in some cases, only vaguely approximate to cycloidal gear design.

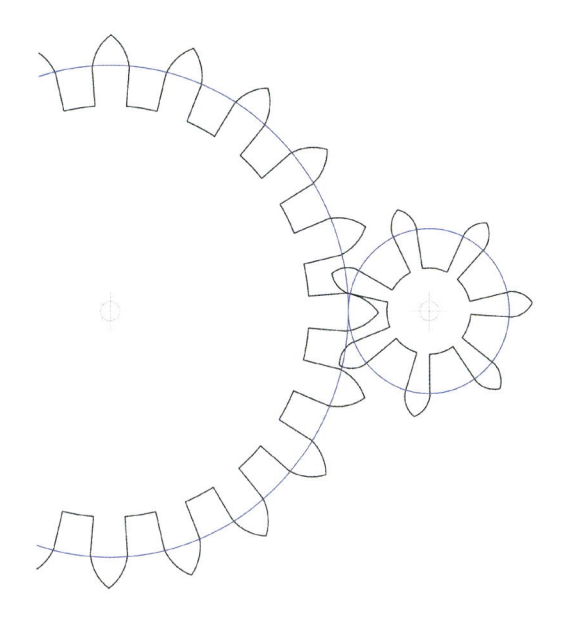

The same arrangement of a twenty-tooth wheel and a seven-leaf pinion, but as a cycloidal form. It is more suited to transmitting power from slower to faster shaft speeds, as in wind- and watermills and clocks.

There are several reasons why early clock wheels are not geometrically correct cycloidal gears. Although the development of longcase clocks commenced at a time of increased curiosity about scientific matters there were practical considerations, and it is quite likely that early clock wheels were made by first cutting slots in the wheel to leave rectangular, flat-topped teeth, and then topping the teeth to leave a more or less triangular top. Thus the shank tends to be radial to the wheel and the top rounded triangular with a flattened top.

There is no doubt that the shape of early longcase clock wheels evolved from medieval or Renaissance clocks, and also from the gearing in windmills and watermills. The clock trade was apparently very conservative and the makers of wheel-cutters learned their skills through the apprentice system without necessarily understanding the geometrical design concepts involved.

Consequently, makers seem to have applied quite crude but individualistic design rules, with the result that the careful clock conserver,

Without the aid of modern manufacturing techniques, true cycloidal teeth (or rather, the relevant cutters) are difficult to make. In the 1950s, about a hundred years after longcase production came to an end and long after European clock manufacturers had embraced the involute tooth form, a British Standard (BS 978 Part 2) was introduced for the benefit of the British clock and watch industry, to formalize some long-standing approximations for cycloidal gear teeth.

The two main features of the BS 978 tooth form are:

- First, below the pitch circle, the shank of the tooth is radial – rather than the theoretical hypocycloidal shape of the lower part of the tooth, it is just cut straight on the radius of the wheel because there is no contact below the pitch circle.
- Second, instead of the difficult cycloidal shape, the curve of the top of the tooth is approximated to a circular arc, the radius of which may be calculated or read off from tables.

The early eighteenth-century hour wheel on the left carries a pinion to drive the date wheel on the right. There is an obvious meshing mismatch and the wheels do not contact on their pitch circles. The lack of uniformity extends to the smaller teeth of the hour wheel with a variation in the cutting depth.

having to make up a replacement wheel, must first understand early tooth designs on a clock-by-clock basis.

Notwithstanding the lack of uniformity across the clock trade during the longcase period, there is often a lack of uniformity of tooth shape within a single movement. While early movements tended to use only one or perhaps two tooth sizes, in later movements the tooth size generally diminishes towards the top of the train.

DESIGNING A NEW WHEEL

There is an over-riding principle that the teeth on any pair of meshing gears should be the same size, pitch and form. Therefore, in a clock repair that involves the replacement of only one wheel, it may be more relevant to base the design of the new wheel on the meshing partner and other wheels within the same movement.

Designing the tooth form of a new wheel really means designing a cutter and the action

Screenshot of a downloadable cycloidal gear design application. The module and numbers of teeth of the wheel/pinion set are entered on the left and the output from the algorithm is shown on the right.

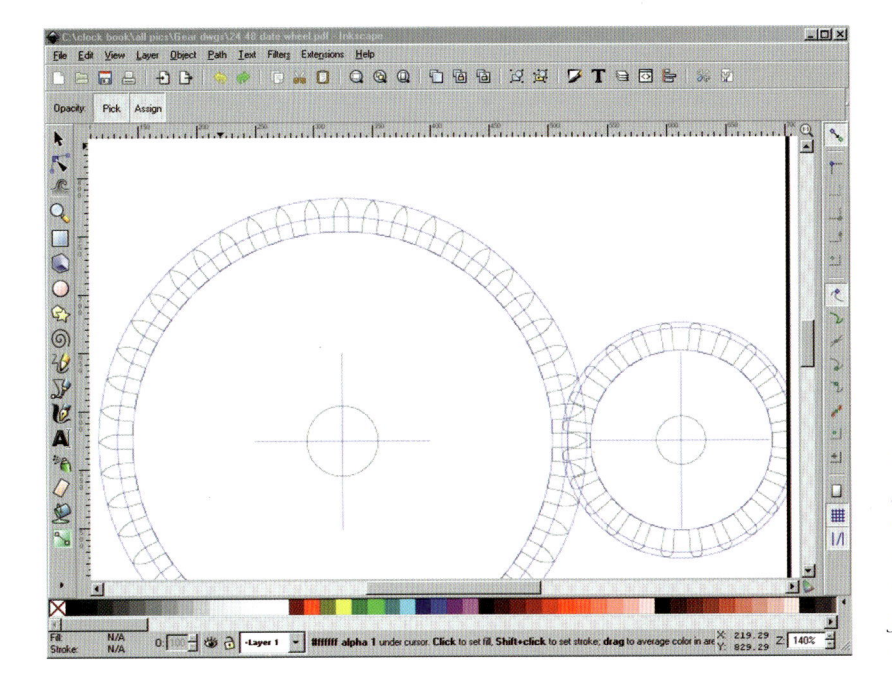

SVG (scaleable vestor graphic) output from the cycloidal gear design application. The image must be converted to a form that can be input to an AutoCAD application.

In addition to the editable and scalable drawing in AutoCAD, a photographic image of a wheel can be inserted for comparison when the shape of the cutter is designed.

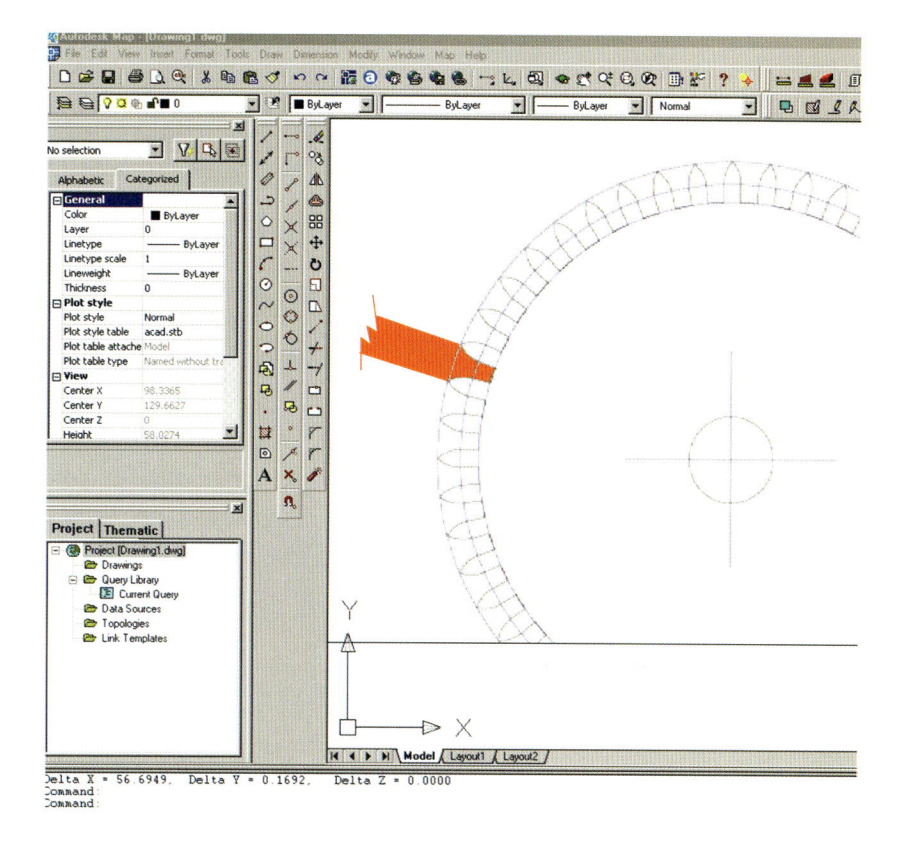

of a gear-cutter is actually to cut away the metal between the teeth. Since the cycloidal tooth form is a reasonably good approximation to the teeth found in longcase clocks, the theoretical cycloidal form is a good starting point in wheel design. For practical purposes, it is worth considering that the design of the tooth form of a new wheel is really about the design of a cutter because making wheels is about cutting away the metal between the teeth.

Drafting an accurate, scaled, design drawing of the new wheel, based on the dimensions of the wheel and the number of teeth, is a relatively easy drawing exercise, but there are various computer techniques that are far easier and more accurate.

Once the design has been completed, there are two approaches to making gear-cutters:

- A simple fly-cutter consists of a single cutting bit set crosswise in a mandrill. It is quick and easy to make and is ideal for one-offs, but is best suited to using in a lathe that has a vertical milling attachment.

- A multi-tooth cutter is more difficult to make and, because each of the cutting edges must have a relief, the cutter should be made of hardened steel to avoid the need for re-sharpening.

A simple fly cutter. An old twist drill or punch is ground to the correct profile and set into the mandrill.

The cutter blank, sawn from an old file, is ready for profiling with the radiused cutter.

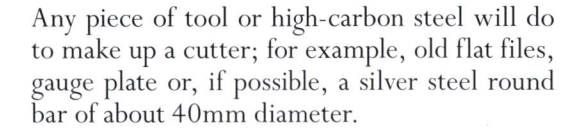

The blank is cleaned up and trued ready for profiling.

Any piece of tool or high-carbon steel will do to make up a cutter; for example, old flat files, gauge plate or, if possible, a silver steel round bar of about 40mm diameter.

Making Up a Multi-Tooth Cutter

Step 1 Make Up the Blank from Suitable Material
Re-using old steel, such as files, necessitates annealing first. The annealed blank is cut out and drilled for a mandrill. After roughly being shaped, it is set up in the lathe and machined to a smooth finish around the edge and across the two faces.

Step 2 Make Up a Radiused Cutter
From the working drawings, the addendum radius of the tooth will be known. A lathe tool is ground to that radius and set up in the tool holder. An easy guide can be made by turning an offcut to the same diameter and using it as a gauge, or even using a twist drill of that diameter.

Only one radius is required because one side of the cutter is formed first and then the cutter blank is reversed in the lathe, so that both sides are made identical. The radius of the cutter will be the radius of the finished tooth.

Step 3 Turn the Cutter Blank to Profile
A straightforward turning procedure that nevertheless takes some careful measurement. It is not crucial that the central part of the cutter is exactly centred but it is easier, especially on a lathe fitted with micrometer indexing on the lateral feed.

The rim of the cutter, which corresponds to the width across the dedendum, is easily measureable; likewise the overall depth.

With the reliefs filed behind the cutting edges, the cutter is ready for hardening.

Step 4 Shape the Cutter Teeth and Form the Relief

Teeth are cut into the profile of the cutter blank and metal is filed away from behind the cutting edges to give them relief (the cutting edge must be more prominent, otherwise they will not cut). At a higher level of sophistication, it is possible, using an eccentric arbor, to machine the relief into the cutter.

Step 5 Harden and Finish

The finished cutter is heated to dull red, quenched in water and polished before heating to straw to red colour to temper. After tempering, the cutting edges are buffed with a fine diamond-coated file.

Using the Cutter to Make the Wheel

Step 1 Preparing the Blank

The thickness of clock wheels varies from 3mm for great wheels down to as little as 1mm for escapement wheels. Once the thickness has been selected, the material should be brought to a suitable temper by light hammering with a planishing hammer on a flat, anvil block.

The centre is marked and a circle scribed before the blank is drilled. The diameter of the blank should be the pitch diameter plus twice the addendum dimension plus an allowance.

The blank is mounted in the lathe and trimmed down to the exact design outside diameter, prior to mounting in the lathe or wheel-cutting engine.

Step 2 Making the Wheel

A lathe fitted with division plates and a vertical milling attachment is ideal for making clock wheels. As an alternative, the homemade wheel

The author's homemade wheel-cutting engine, based on early designs but with an electric motor (cover removed for illustration).

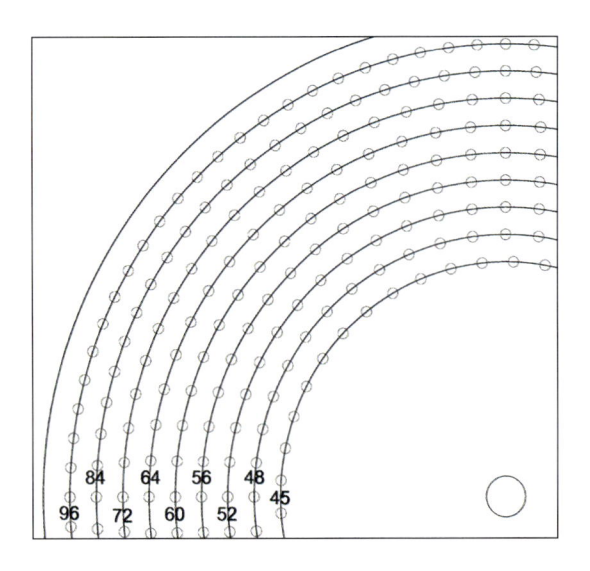

In AutoCAD, the Divide command is used to insert a specified number of points evenly on each of a set of concentric circles.

Another view of the wheel-cutting engine showing the indexing pin and the rings of holes in the dividing plate.

engine illustrated is relatively easy to make up using a small electric motor, either 12V such as a car window winder or windscreen wiper motor, or a small domestic appliance motor.

The most accurate cutting is done by an up and down, slicing motion of the cutter through the wheel blank. That method ensures an even depth and symmetrical teeth.

An essential part of the wheel engine is the dividing plate and, again, modern technology is useful. Marking up and making dividing circles used to be a matter of carefully measuring chords that were calculated from mathematical tables. With access to computers, a paper template can be printed after only a few key-strokes.

The printed paper template is then pasted to the blank dividing circle, which is punched at the marks and drilled.

Crossing Out

Traditionally, wheels inside the movement plate are crossed out to remove as much weight as possible. Outside the movement plates, the motion-work and date-work are rarely crossed out.

The wheel is set up in a lathe to scribe the outer limit of the crossings and then on the bench, the spokes are scribed with a sharp graver.

As an alternative to the traditional method of marking out, the crossing can be drawn up in AutoCAD, and the paper print pasted to the wheel.

An off-the-shelf multi-tooth cutter in the wheel-cutting engine. The blank should be well-supported underneath to prevent distortion.

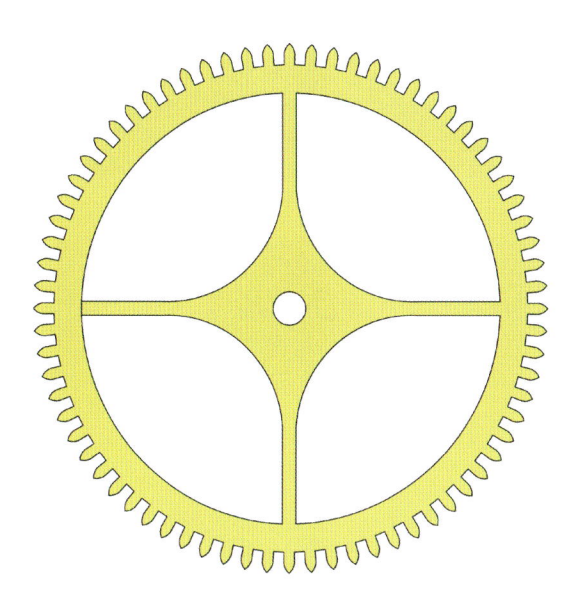

Crossing out is done to a very specific design.

This 'odd' 30-hour movement is obviously early and, in common with many early clocks, all of the teeth are the same size.

A Note About the Suitability of Off-the-Shelf Cutters

Unfortunately, cycloidal cutters are now as rare as the wheel-cutting engines that clockmakers once used. However, they are available from several manufacturers; they come not singly but as a set of about eight because the tooth shape is different depending on the number of teeth in the wheel. Consequently, a set is required for every different tooth size, so a complete set of cycloidal cutters is likely to involve substantial cost and is probably more for the serious professional.

Designing New Motion-Work – an Example About Expecting the Unexpected

Whether damaged through excessive wear or by broken teeth, it is occasionally necessary to make up replacement wheels. The procedure set out below is about the replacement of a missing motion-work. The clock in question had been acquired in an auction and the buyer was led to believe that, regardless of the missing dial, there was 'something special' about the movement.

A standard 30-hour movement showing the motion-work.

Calculating the tooth count

	Typical 30-hour movement (with a 1sec pendulum)	Example 'odd' movement (assuming 1sec pendulum)
Escape wheel	30	32
Escape wheel pinion	6	6
3rd wheel	72	72
3rd wheel pinion	6	6
Great wheel	90	84
Minutes for one revolution of great wheel	180	179.2

The movement is obviously quite old and a repairer's name is scribed in the inside of the back-plate with the date, 18 October 1808. Since the motion-work was completely absent, a decision was made to make the replacement wheels in order to return the clock to working condition.

The first step in making up the replacement wheels was to calculate the tooth count, i.e. the reduction from the pinion of report on the great wheel arbor to the hour and minute wheels.

Compared to a typical 30-hour, three-wheel movement, the tooth count was a little odd and, at first sight, nonsensical.

Calculation of the required tooth counts on the new wheels that were required to drive the hour and minute hands yielded very large numbers of teeth to get the gearing right until, in a Eureka moment, it was realized that this movement had been made for 1¼ sec pendulum. Thus the great wheel arbor turns once in 1¼ × 179.2 = 224min. So the ratio from that to the hour wheel resolved to fourteen teeth to forty-five, and the ratio to the minute wheel is fifty-six to fifteen.

Measure and Calculate the Required Wheel Diameters

The measurement between the centre of the great wheel arbor and the centre of the mounting post for the hands is equal to the sum of the pitch radii of two meshing wheels.

Since the size of teeth in each pair of meshing wheels must be the same, the ratio of the wheel diameters is equal to the tooth ratio.

Distance between centres is 37.7mm.
The hour wheel ratio is 14:45.
Hence, hour wheel (pitch) diameters are:

2 × 37.7 × 14/(14 + 45)mm = 17.9mm (driver).
2 × 37.7 × 45/(14 + 45)mm = 57.5mm (driven).

Hence the module is 1.28.
The minute wheel ratio is 56:15.
Hence minute wheel (pitch) diameters are:

2 × 37.7 × 56/(15 + 56)mm = 59.48mm (driver).
2 × 37.7 × 15/(15 + 56)mm = 15.92mm (driven).

Hence the module is 1.06.

The diameters given above are the diameters of the gear pitch circles, not the overall diameter of the wheels. Overall diameter (the diameter of the blank) is equal to the pitch diameter plus twice the addendum. It will be seen that, because the total number of teeth on each meshing pair is different, all the teeth are different, either in size or shape. While meshing teeth are the same size, measured in module, the shape depends on the number of teeth on the wheel. Consequently, four cutters will be needed, one for each wheel.

The wheels are cut as described above and once the teeth are cut, the burs are removed and the teeth cleaned up with fine emery.

Mounting the Wheel

In the example shown, the driven wheels are locked together, tightly fitted and soft soldered onto a brass tube that fits over the extension of the great wheel arbor. The driven wheels are mounted on their separate tubes, which run concentrically on a steel post screwed to the front movement plate. The tubes are turned up with a shoulder to support the wheel.

In designing the tubes for the driven wheels, consideration should be given to the size and shape of the hands and the way it is intended to mount them. Early steel hour hands tend to have large, cut squares and fit tightly over the square end to the hour wheel tube. Later types are usually cut circular, with a small fixing screw to hold the hand to the stepped tube.

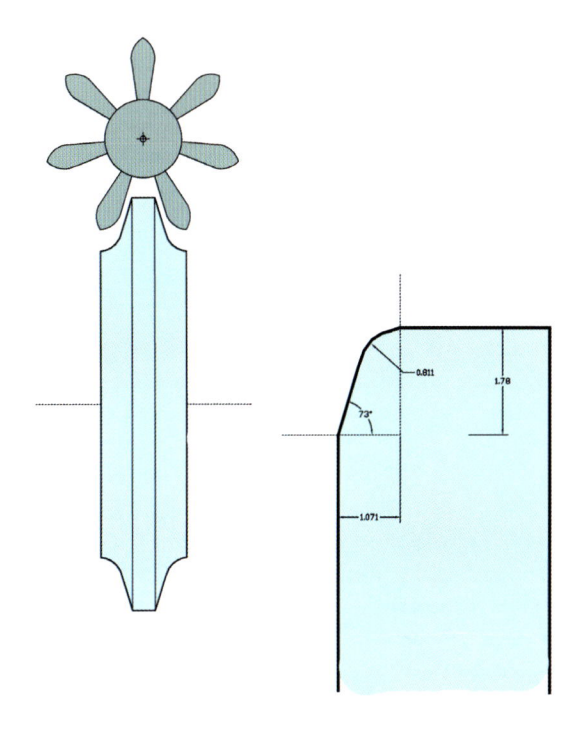

MAKING A REPLACEMENT ARBOR AND PINION

Unlike brass wheels, which are fixed to their arbors, either riveted silver soldered or, in later clocks, by soft soldering onto brass collets, pinions are part of the arbor and cut in the arbor blank as it is formed.

The pinion cutter (left) is designed from the standardized rules set out in BS 978 Part 2 but care should be taken to ensure that this theoretical pinion leaf shape is acceptably close to the actual leaves in the clock movement. The lathe tool for cutting the pinion cutter profile is shown on the left scaled up by ×2.

This pinion has worn far enough to require either moving the meshing wheel along its arbor to present a new wearing surface onto the pinion or, otherwise, a complete arbor and pinion replacement.

Although the material is hardened high-carbon steel, wear can develop in the pinion leaves, gradually wearing them away until breakage is likely. Replacement of an arbor may become necessary if one of the pivots is worn thin or broken off. Longcase pinions are almost always of six, seven or eight leaves; the tooth shape is invariably some form of cycloidal with radial shanks.

There are two approaches to forming the arbor and pinion:

- Where a suitable milling machine is available, including a pinion cutter of the correct profile, the pinion form is cut over a relatively long length of round bar, which is turned down to the arbor diameter, leaving the relatively short pinion. The workshop procedure should not be particularly challenging to a skilled machinist. A pinion cutter is made up from high-carbon steel, as previously described, from the known geometry of the required pinion.

- The pinion leaves are hand sawn and filed from the blank in a pinion filing jig. With a little practice, quite satisfactory results are possible, but the procedure (set out below) relies on a (probably) homemade jig.

In each case, the steel blank is annealed prior to cutting and then heat-hardened after.

Hand-Cutting

Step 1 Make the Blank Arbor

Using the old pinion as a pattern is the preferred approach to setting out the replacement. An alternative method is to estimate the pitch diameter and from that, the outside diameter of the new pinion blank.

The pinion is usually located at one end or other of the arbor, and the easiest procedure is to first reduce the diameter of the silver steel round bar to that of the intended pinion size and then shape the pivot and the end of the arbor opposite to the pinion. The blank is reduced as far as the inner face of the pinion. Reversing the blank in the lathe, the outer face of the pinion is formed and then the short end of the arbor and finally the pivot.

Step 2 Set Up the Blank in the Pinion Jig and Make Initial Cuts to the Depth of the Tooth Gaps

A dividing plate is an absolutely essential element of the pinion jig, as is the ability to make the saw-cuts true (i.e. with the saw blade centred over the axis of the arbor and the blade vertical).

Where possible, the saw-cuts may be made with a wider blade and, ideally, the saw-cut will be the width of the design gap at the tooth root.

A simple pinion-cutting jig contains a dividing circle with rings of six, seven and eight holes.

The stages of cutting a pinion by hand. Top left: radial saw-cuts are made to the full depth. Top right: the saw-cuts are opened up with a knife file. Bottom left: the addendum is roughed out with a square file. Bottom right: the leaves are shaped with a fine, flat file.

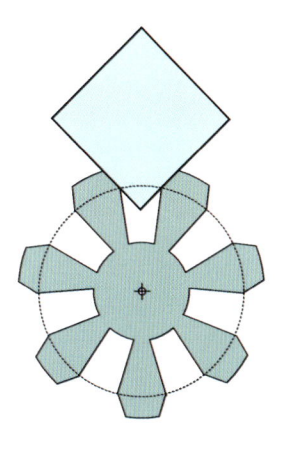

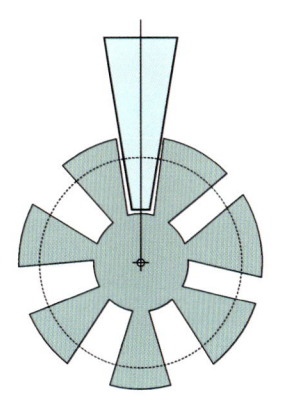

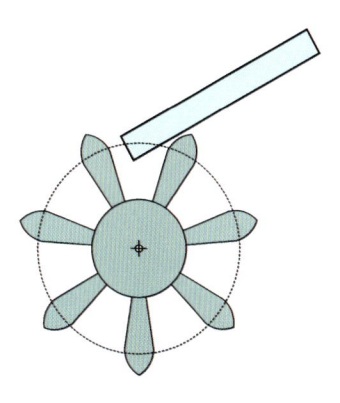

Steady strokes with a sharp saw will keep the cut straight and true.

Step 3 File the Addendum Part of the Leaves

The addendum approximates to a circular arc and may be roughed out using a variety of files, first triangular, then square and then flat. At each stage the work is compared to the original for overall leaf shape. When the shaping has progressed sufficiently, the new pinion is tried in the depthing tool to give an idea of smoothness of operation.

With practice, and by repeatedly examining and reworking the pinion leaves, a completely acceptable replacement can be achieved.

Step 4 Harden, Temper and Finish the Arbor

The finished arbor is wound in binding wire, which distributes the heat more evenly, and then, when heated to a deep cherry red in a gas flame, it is quenched in water. It is important

Final filing is best done with the aid of an eyeglass to ensure that the leaves are uniform. (It takes practice.)

to quench the entire piece together to prevent distortions.

If the hardening process has been done correctly, the dark outer layer will scale off to show a white steel surface. A traditional way of tempering to reduce the brittleness of the quench-hardened steel is by first binding in wire and dipping into heavy oil before placing in a gas flame. Once the oil starts to release black smoke the temperature will be high enough, and the tempering process is stopped by plunging into water.

The arbor is tested for straightness between lathe centres using a piece of chalk, which will mark a high spot; the chalk mark is then tapped with a cross peen hammer on a steel stake.

The shoulders and pinion leaves are buffed with emery and then polished and the pivots burnished before trying the completed arbor in the depthing tool and finally, fitting to the clock.

UNDERSTANDING RACK GEOMETRY

Snags in the striking cycle develop through damage, wear or lost adjustment of the rack and its associated parts; the result is miss-counts or complete absence of striking.

If the rack fails to trip for any reason, the rack-tail pin will collide with the step in the hour-wheel snail sometime after twelve o'clock. To protect the clock from damage or stalling, the rack tail is designed so that the pin skips over the step in the snail as it collides.

When the rack-tail is adjusted correctly, the rack-hook should just fall cleanly into the appropriate gap between rack teeth at the commencement of the strike cycle. At each revolution of the gathering pallet, the rack should advance by just one tooth. Although there is some tolerance, if the gathering pallet is too long it will tend to over-reach, putting more wear on the rack hook. If the gathering pallet is too short, there will be a possibility of the rack hook failing to step over the tooth, with the result that the strike will continue indefinitely.

Occasionally a substitute rack is introduced into the clock and, if the pitch is different to that of the original, which is quite likely, the strike cycle will be problematic.

Making a new rack is not necessarily difficult but the setting out tends to cause problems.

There are two dimensions to consider when designing a replacement rack:

- First, because the rack teeth are arranged in a curve, that radius is governed by the distance from the rack post to the gathering pallet, so that at any position of the rack, the gathering pallet will mesh to sufficient depth to draw the rack forward by one tooth.
- Second, the overall length of the rack from first to twelfth tooth is governed by the ratio of the length of the rack from post to teeth to the length of the rack tail and also the drop of the steps of the snail.

Making Up a New Rack

Step 1 Design

If CAD drawing software is available, set out the salient geometry of the hour-wheel snail, the rack mounting post and the lowest point of the gathering pallet during its rotation. Otherwise, draw by hand, not forgetting the position and dimensions of the gathering pallet.

The rack mounting post is arranged so that the length of the rack-tail from centre to pin

The leverage effect in the geometry of the rack means that the throw of the rack teeth is 'geared up' by the ratio of the rack radius to the length of the rack tail.

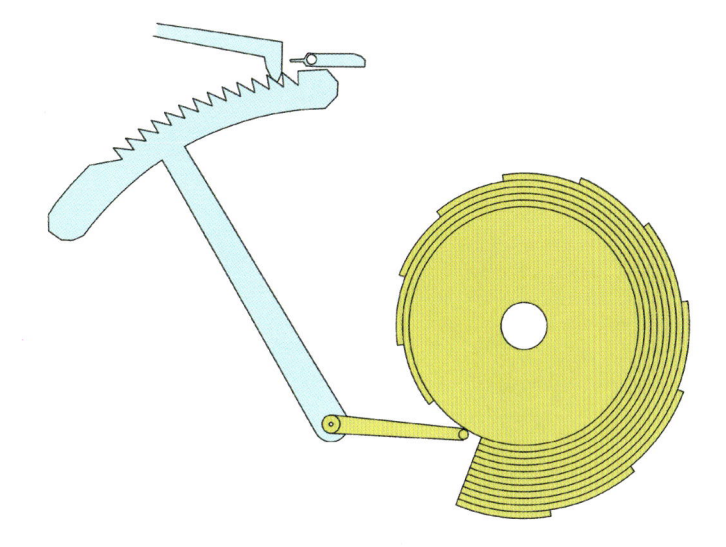

is equal to the distance from the rack post to the hour wheel, which will ensure that the pin moves more or less on the hour wheel radius.

The rack tooth pitch is calculated by measuring the radial drop from twelve to one on the snail, dividing by eleven and then multiplying by the ratio of the arm length to the tail length. The depth and shape of the rack teeth will reflect the shape of the rack-hook pawl and the gathering pallet.

Step 2 Mark and Cut Out the Rack

The rack is made up from a piece of mild steel about 2mm thickness. The rack teeth are scribed onto the steel blank or, alternatively, a paper pattern is glued onto the steel as a guide. It is helpful to accurately shape the arc described by the top of the rack teeth before attempting to cut the teeth. The rack is cut out rough with a fine-tooth saw, filed to final shape and, finally, buffed with emery and polished.

The hole for the mounting pipe is drilled and a pin is fitted that limits the gathering action by stopping the blade at its last rotation of the strike cycle. Careful measurement of the rack tail is needed to ensure that the gathering pallet is stopped by the pin on the last revolution of the strike cycle.

A hook is cut for the end of the rack spring and the new rack is quench-hardened and then tempered to straw colour, so that it will be reasonably resistant to bending.

PAINTED DIALS

The introduction of painted dials in the 1770s brought longcase clocks within the budget of a large sector of the population and, consequently, a very high proportion of longcase clocks date from that painted-dial period.

The chip of paint lost from near the Jay's tail coincides with the rivet that fixes the dial foot. Paint losses are very difficult to repair but, if possible, any other loose chips should be stabilized by introducing a liquid glue between the base of the chip and the steel dial.

Unsightly rust and flaking paint. This dial is probably too good to destroy by restoration and it is possible that a little careful work would stabilize any loose paint and perhaps tone down the appearance of the losses.

The maker's name (John Fisher, Bilston) is only legible with an ultraviolet light. The seconds marks have all but disappeared and the hour marks are characteristically rubbed, where a finger has pushed the minute hand around to reset the time. This dial might benefit from some gentle restoration to replace some of the lost inking without damaging the overall mellow look.

Painted dials are prone to a number of faults:

- Because the dial feet were fixed before the white paint was applied, they are vulnerable to stresses and impacts, which tend to throw the paint cover from the dial foot rivet.
- Being iron, they are prone to rust, which causes the paint to flake off.
- The hour and minute marks and the maker's name were applied in black ink, which is prone to rubbing off.

The conservation principle of minimum intervention is often in conflict with aesthetic considerations in various aspects of longcase clocks and none more than painted dials. While it might be good conservation practice to leave a badly degraded dial, most owners prefer a more radical restoration, even renovation to an as-new look.

In extreme cases there is no choice but to completely repaint, but even then, there are usually traces of the original paint and a detailed

Too far gone to save, this dial looks like an ideal candidate for practising the skills involved in dial restoration. But, before stripping off the remaining paint, the detail should be carefully recorded.

The Most Difficult Workshop Procedures   143

photographic record should be taken so that the dial can be repainted in the original style.

Once the detail has been recorded, the old paint can be removed and the metal surfaces cleaned and primed. Good-quality enamel paint is applied and baked to give a hard, durable finish.

Lines and rings are drawn using drafting pens, while lettering is best applied with old-fashioned dip pens. A good-quality Indian ink is used and, because it is intended to be applied by pen onto paper or drafting film, it is too watery, so it should be evaporated to about 70 per cent of original volume first and mixed with some soap to let it fix onto the paint surface.

Re-applying paint is a skill that takes years of practice to acquire, so the only circumstances in which a complete novice would be justified in attempting to re-paint a dial are when the dial is a complete write-off. For most painted-dial restoration work, the skills of a specialist restorer are the preferred option, but for the enthusiastic novice, practice is the key element and there are no short-cuts.

Very close examination of the dial shown opposite reveals a painted vignette of Saint George slaying the dragon.

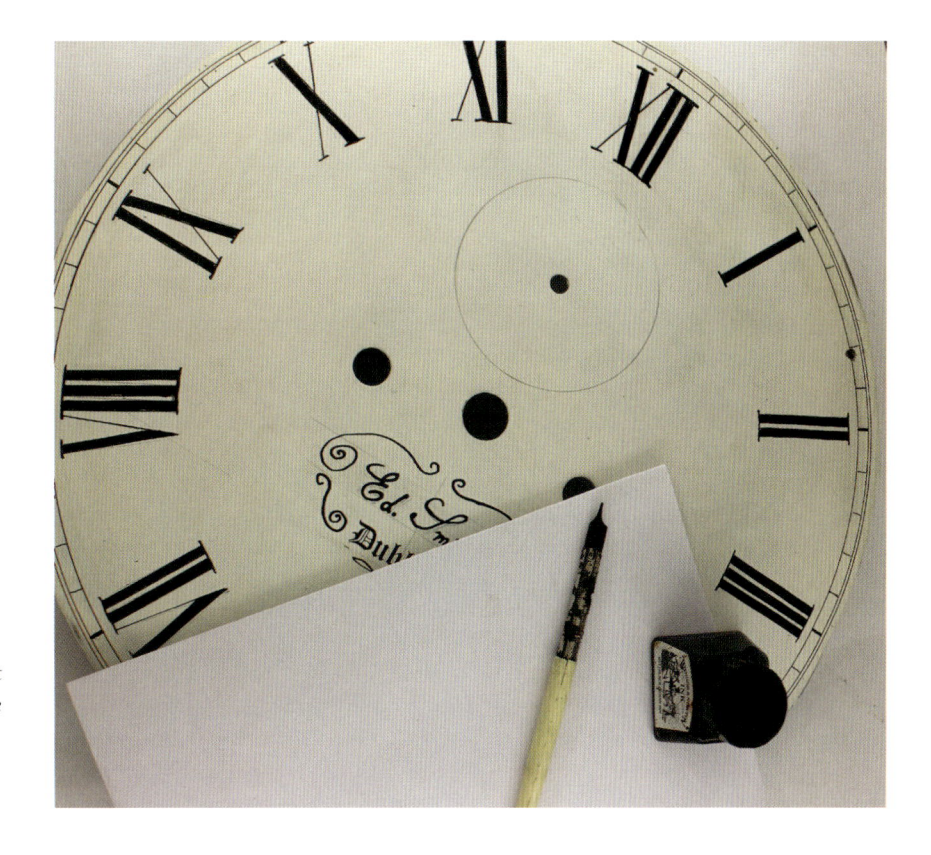

The delicate work of applying ink is best done with dip pens, which come with a range of nibs. If the inking is not correct, it is easy to remove before it dries. Once finished, it should be baked to properly dry.

Chapter 8

Case-Work – Maintenance and Simple but Effective Repairs

The general arrangement of case construction is outlined in Chapter 1 but, as with every other aspect of longcase clocks, the exceptions are frequent. In this chapter, some of the more superficial restoration procedures are described: wood finishes, veneers, inlays, mouldings, hinges and locks.

The concepts of careful restoration and conservation of historical artefacts apply equally to clock-cases as to their clocks, so if it is decided that the structure or finish has deteriorated enough to merit some repair work, the first step is to examine the whole case carefully and identify the various structural elements and how they are joined; also the type and condition of the veneers, inlays and finishes used. Usually a small, inconspicuous area can be found (typically lower back on the side of the base or an area that has sustained heavy damage), where a small area of the surface can be carefully scraped back to test for separate layers of different finishes.

Conservation is about stabilization and preservation for posterity. A professional conserver, whether he is engaged in private clients' work or in the museum environment, follows a rigorous procedure that commences with a condition report upon which some strategy is decided.

By comparison, the concepts and practice of restoration are far more subjective because it is about keeping or restoring functionality in the most sympathetic way possible. Consequently, the restorer's role is more difficult than that of the conservator; he is required to make aesthetic and practical judgements about how to proceed with the work, including choices of materials to be incorporated in new work, while keeping the character of the original.

The amateur clock-owner/restorer is strongly advised to follow the same preliminary procedures, rather than rushing into an unplanned programme of repair work, which can often result in a series of isolated mends and a generally untidy restoration.

A WORD ABOUT GLUES AND GLUING

The shelves of a modern hardware store are filled with dozens of different types of glues and adhesives. There is a wide range of chemical adhesives, some of which are suitable for restoration, but what is rarely seen on those shelves is the type of glue that woodworkers used in the construction of the cases for longcase clocks. It is often called scotch glue or animal glue and, before the modern age, very effective glues were made from the hooves and horns of animals, fish bones, rabbit skins, etc. The adhesive properties are based on the protein, collagen, which is found in connective tissue in mammals; it has the advantage that the gluing action is reversible, with the application of heat. The solvent is water, which tends to lift wood fibres as it absorbs into the gluing surfaces, strengthening the bond.

Traditional scotch glue is available from specialist suppliers and is the preferred material for conservation because it is likely to be as close to the original as possible.

In restoration work, to make a satisfactory repair, the old glue should be removed using very hot water or steam, so that the new glue can make a firm mechanical bond with the wood fibres. Once the re-assembled parts have been glued up, they should be clamped tightly to make the glue-joint between the parts as thin as possible. As a general rule, the thinner the glue line, the stronger the joint, but clamping the joint together while the glue sets also exerts a pressure that forces glue into the wood fibres. These basic principles apply to the use of all glues, whether traditional or modern.

MAKING UP SCOTCH OR ANIMAL GLUES

Scotch glue is obtainable in the form of powder or beads. It is soaked in water and heated to just 60°C (hand-hot) in a glue-pot. When it is mixed correctly it runs in a thin thread, not in drips. It is applied and spread hot, so on large surfaces it is a good idea to pre-heat the work before spreading the glue and even then, work quickly. The adhesive properties deteriorate if it is reheated, so only enough should be mixed for one gluing session. Once glued, the work should be clamped up and left for several hours (conveniently, overnight). Animal glues are thermo-softening, which makes them ideal for veneer work. Preparing and using animal glue

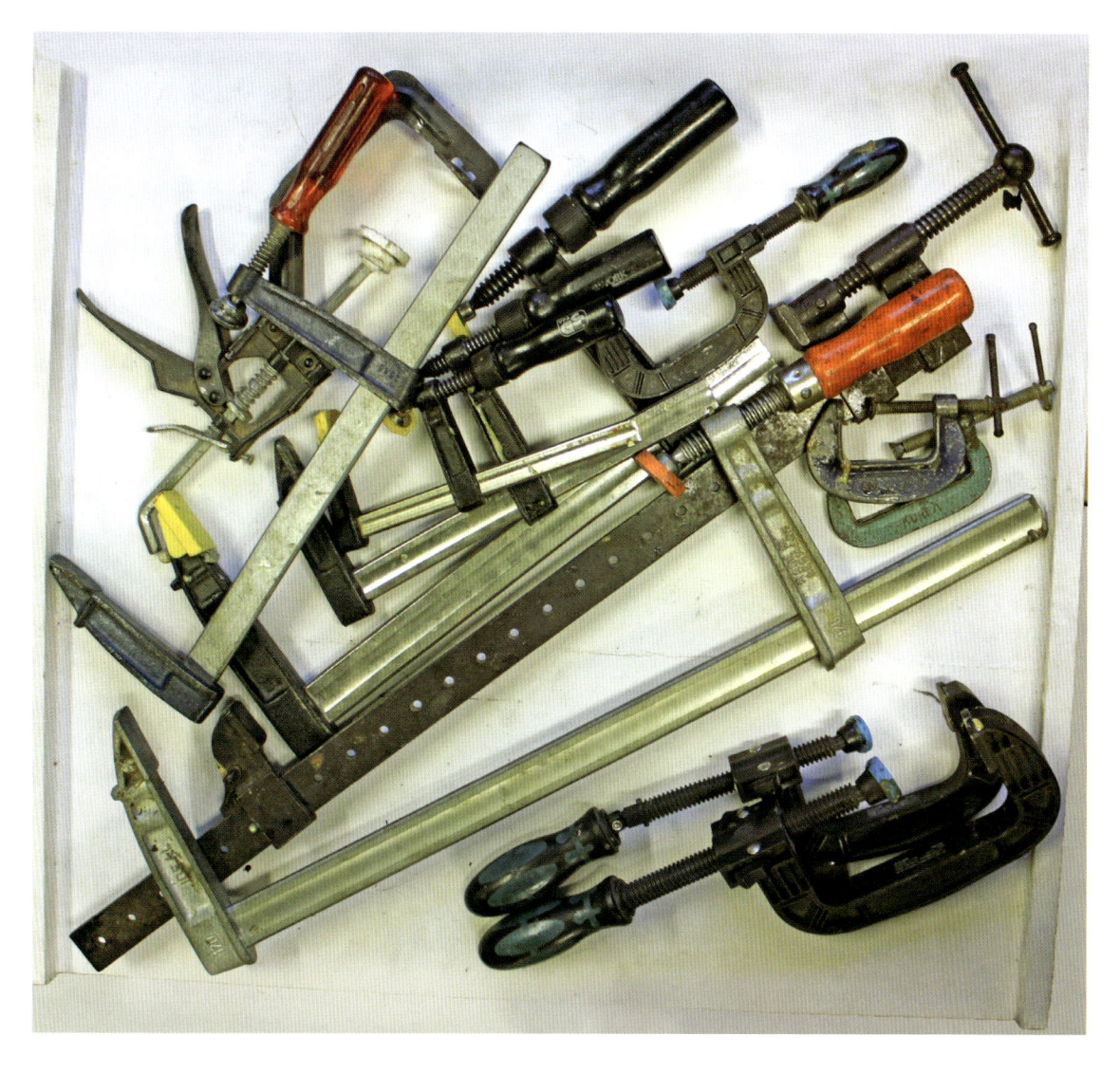

Absolutely essential — no successful gluing without effective clamps. They come in all sorts of shapes and sizes.

takes practice, but once it has been mastered it is probably the best all-round glue for antique woodwork restoration and conservation.

OTHER TYPES OF GLUE

PVA Wood Adhesives

An emulsion of polymerized vinyl acetate in water with added chemicals that modify viscosity and plasticity. It is a useful cold-working substitute for animal glues and does not stain veneers, but is susceptible to damp and requires firm clamping of the joining parts. It is not thermo-softening.

Aliphatic Resin Adhesives

Similar to PVA adhesives but preferable for some applications, especially restoration and veneer work. They are thermo-softening, so once applied to veneers they can be re-mobilized by applying an ordinary domestic smoothing iron. For applications where waterproofing is not a necessity, aliphatic resins have mostly replaced urea formaldehyde glues, such as Cascamite, that appeared in the 1950s and 1960s.

Glue-Film (Hot-Melt Film)

Hot-melt film is a type of polythene sheet. It is only suitable for veneer work and is supplied in sheets with a backing paper. It has an indefinite shelf-life and is clean and easy to use. It has several advantages over applied wet glues: when it is ironed onto veneer pieces it reduces the risk of small patches breaking up during cutting and preparation; it also allows different veneers to be assembled dry on the work and then glued together. (When water-based adhesives are used with veneers, there is always a likelihood of shrinkage as the moisture dries out.)

Thermo-Softening Glue Sticks

The same adhesive chemistry as hot-melt film is available in stick form and is widely used in the craft-work and DIY sectors. Sticks of the hot-

Scotch glue is the preferred option for conservation, except where stability could be improved by using a modern glue. The aliphatic resin glue (left of centre) is very similar to animal glues. Hot melt film and sticks (top) with an electric hot glue gun.

melt adhesive are heated in an electrically heated glue gun. It is ideal for small veneer patches.

Epoxy Resins

The components of two-part epoxy adhesives are mixed just prior to application. They are expensive and only moderately suitable for woodwork repairs (but not conservation). They are not particularly good for filling gaps without the addition of special extenders. They are not suitable for conservation work and are difficult to remove.

Polyurethane Adhesives

A relatively new innovation, polyurethane glues are supplied as viscous liquids or gels, in bottles or mastic tubes. During the curing process, which is accelerated by the presence of moisture, the adhesive foams, thereby creating an internal pressure within the joint if it is securely clamped; otherwise there is a tendency for the expanding foam to force the joint apart. It is most useful for wood-to-wood joints in general wood repair-work and restoration, but as a material for conservation its use should be weighed against more traditional alternatives, especially scotch glue. Its advantage in restoration work is in the excellent gap-filling that makes up for gradual erosion of wood fibres from loose construction joints. Polyurethane adhesives tend to be messy and their main disadvantage in antique woodwork repair can be the foaming extrusion of excess glue from the curing joint. If it is allowed to come into contact with fresh wood, excess glue will prevent the absorption of stains and dyes, so as a precaution, an application of shellac will prevent absorption and aid the removal of excess glue after it has cured.

WOOD FINISHES

Clock-cases were made from all sorts of woods and, occasionally, covered with veneers from exotic or decorative woods. A special form of veneering, marquetry, was introduced to English furniture and clocks from Holland in the late seventeenth century. A variety of finishes was applied to the wood cases of longcase clocks: waxes and oils, lacquers, spirit varnishes,

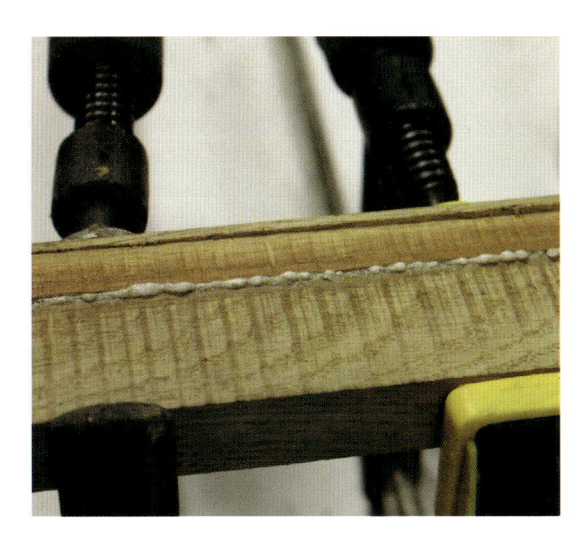

Polyurethane glues always foam out as they cure, so good clamping is vital.

The spoiled appearance of this mahogany cross-banded oak door has been caused by removal of the top surface of the oak during restoration of the cross-band. Oak darkens with age and sanding has left a bleached look, which might have been saved by treating the bared oak with potassium permanganate.

A victim of a former fashion for stripped pine. Pine clock-cases with their original painted finish that imitated exotic woods are now very rare.

Softwood clock-cases, especially later, low-cost country-made examples, were painted in imitation of exotic woods and quite often embellished with a wood-grain effect. Sadly, the 1960s fashion for stripping painted pine furniture has persisted to the present day to the extent that it is now quite rare to find a painted softwood clock-case.

The basic rules for working with the finishes on clock-cases are the same as in antique furniture:

- First, decide on the rarity of the piece and if it is to be conserved (i.e. stabilized and preserved) or restored to a supposed condition sometime in its past, possibly, but not necessarily as it was when new.
- Never use sandpaper to remove finishes – the effect is uneven, and it damages the edges and removes the surface of the wood that has darkened over the years. The effect of sanding mahogany can be quite alarming because it was originally darkened from the salmon pink colour of its raw state by applying potassium dichromate or some other strong agent, which only soaks into a thin surface layer.
- Stripping off the original finish with chemical paint-strippers is absolutely the last resort – sometimes it is necessary, as, for example, when a modern varnish has been brushed on roughly or, worse, sprayed on.
- Removing the surface patina from wood finishes should be avoided but there is a difference between the effect of rubbing and polishing over the years (true patina) and plain grime. Patina is a term more applicable to wax finishes, when the appearance is more of burnished wood. Shellac polish naturally deteriorates, especially in strong sunlight to an opaque greyish yellow and it is perfectly acceptable to replace a shellac-polished finish periodically.
- Only remove as much of the finish as is necessary to allow the colour and figure of the wood to show. In other words, when removing discoloured shellac polish, it is not always necessary to dissolve away all of the polish, just enough of the opaque top surface until the wood shows through.

shellac polishes and paint. To clarify the two terms, French polish and shellac polish, which tend to be used synonymously, shellac is the material that is dissolved in alcohol to make a shellac polish and French polishing is the process by which it is applied to wood surfaces. So, a French polished case is one that has been polished by that method (described later) using shellac polish.

Unfortunately, restoring the finish is not just a simple matter of working with any one type of finish because over the years, the appearance of clock-cases was often improved by applying another finish. It is, therefore, not unusual to find a modern varnish over a degraded French polish over the original wax.

- Never fill dents and holes in fine woods with inert opaque materials, such as car body fillers, prior to applying a replacement finish. They can never be coloured to match because the appearance of woods changes with lighting and viewpoint. Hard waxes and solid shellac are more translucent and can be used to fill holes but the result will always show.

IDENTIFYING FINISHES

Appearance

When the appearance of the finish is shiny or glazed, with little or no open-pore structure in the wood surface, the finish is quite likely shellac polish, but identifying shellac polish is easier if the wood type and overall style are considered. As a general rule, the exotics are polished, while the native woods are waxed. In addition, French polish is more appropriate to the darker woods and does not particularly enhance the lighter shades. The situation is further confused by the relatively late introduction of French polish into English furniture about the time of the French Revolution. Although shellac had been known and used in France and Holland for many decades, once introduced into England it rapidly spread and pieces of domestic furniture (including clock-cases) that had originally been burnished with wax were re-finished in French polish in the new fashion.

Consequently, one might expect a mid- to late Georgian formal clock-case, typically of mahogany, to be finished in a shellac polish laid onto the original wax, while its country counterpart might retain a wax finish to its plain oak case.

Solvent Testing

A useful preliminary test to identify shellac polish relies on its solubility in spirit (either ordinary ethyl alcohol or industrial methylated spirits). The surface is first wetted with a small swab containing a few drops of the spirit. The spirit evaporates readily and after a minute or two it is rubbed more vigorously with the swab. Shellac polish will soften and dissolve in

The surface is moistened with spirit (alcohol) and rubbed lightly with a fine wire wool. Shellac polish will lift off, to give a dark smudge on the swab.

the spirit, leaving it sticky to the touch; it will leave a sticky deposit on the swab.

If the spirit does not soften or dissolve the finish it is unlikely to be a simple shellac polish, so the process is repeated with turpentine, which will soften or dissolve early types of varnish and wax.

FRENCH POLISHING (APPLYING SHELLAC POLISH) – SOME GUIDANCE NOTES

French polishing takes practice and patience and the novice can get into all sorts of problems, but shellac polish is very forgiving and, with some perseverance, perfect results can be achieved quite quickly.

- The rubber or fad, as polishers used to call it, is made up to suit the size of the work. It may be necessary to fold the fabric of the rubber to a reasonably well-defined (but not sharp) point in order to reach into angles and mouldings. The best sort of fabric to use is an old T-shirt, but any 100 per cent cotton will do. For large areas, the sides and door of the trunk and the base, a ball of shredded cotton

A square of cotton fabric is folded across the diagonal. Shredded cotton waste is ideal as stuffing for a rubber.

The point and opposite edge are pulled tightly over the stuffing.

Each of the tails is twisted to hold the stuffing in place.

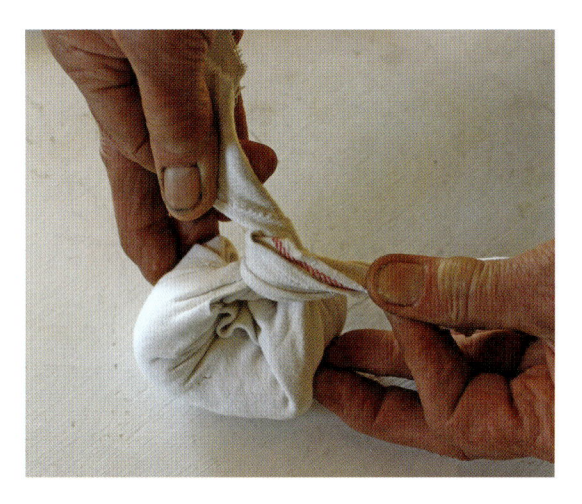

The two tails are twisted together to make a handle.

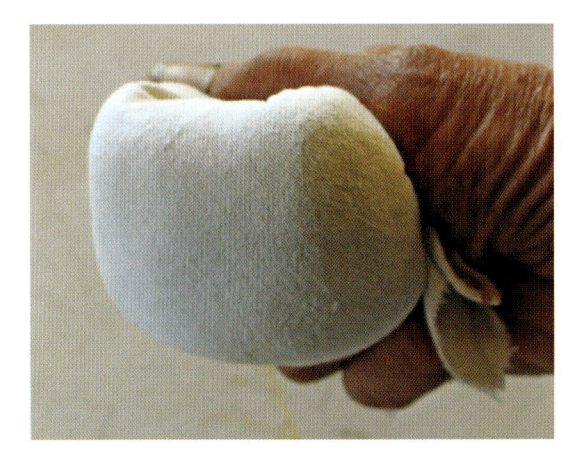

The finished rubber should be firm but not hard. If the fabric is loose, it will wrinkle and mark the polish.

rag is placed inside the folded rubber to act as a reservoir for the polish. For smaller work, it is easier to use a rubber folded from several thicknesses of fabric.

- The polish should be applied thinly and never wet; when the rubber has been charged with polish, the surface should just seem moist. A good way of testing a freshly charged rubber is to press it firmly onto a clean sheet of ordinary copy paper. The resulting mark should not be a wet blob; it should just show the weave of the fabric.
- The polish should transfer from the rubber to the work and as the rubber dries it may start to stick to the work. A drop of boiled

linseed oil will keep the rubber moving, so that, with increasing pressure, it will deposit more polish and smooth the surface of the work.
- Shellac polish is its own solvent and when the rubber is recharged with polish there is a risk of marking the polish that was previously laid onto the work. The rubber should be used lightly and briskly at first; success at this stage is to do with the polisher's 'feel'.
- On flat surfaces, the polish is laid on by the overlapping circular movement of the rubber and then the built up polish is worked down flat using long sweeps of the rubber, with plenty of pressure on the rubber.
- Always work away from stops or internal angles in the work because, if the rubber is allowed to stop, even momentarily, on the work, it is likely to stick and leave a mark.
- After about three rubs of polish, the work may be set aside for a while to harden out slightly and then cut back, either with fine wire wool or, on flat surfaces, a fine wet and dry paper (about 400 grit) with the surface wetted with turpentine. Wire wool conducts heat away from the area being rubbed and likewise turpentine prevents a build-up of heat, but in each case there is a risk that the fine shellac dust liberated will coagulate and stick to the work, so the rubbing should be done slowly without excessive pressure.
- The role of linseed oil is often misunderstood; it forms no part of the final polish and is used only to lubricate the rubber, so as to extract extremely thin layers of polish from a quite dry rubber. Once the work has begun to harden out, over a matter of days, the oil will appear on the surface and it should be removed with a clean cloth moistened with turpentine.
- The polished surface is burnished using a pad of cotton onto which one or two drops of spirit have been thoroughly distributed. The pad is initially worked across the surface lightly and briskly, and with increasing pressure in a circular motion. Any marks on the surface will flatten and the whole surface will develop a flat, mirror-like finish.
- Certain very old woods develop a corrugated surface as the wood shrinks back on the grain and that finish should not be flattened out by polishing.
- The objective of French polishing is to give to the wood surface the look of having been burnished with wax.

DETACHED GLUE-BLOCKS

Softwood glue-blocks are used extensively in clock-case construction; they enhance the stiffness of right-angle joints between adjacent sheets of wood.

Loss of glue blocks through dampness or rough handling renders the entire case vulnerable to serious damage, so it is unwise to overlook a few loose blocks in the bottom of the case.

The softwoods normally used for making glue-blocks usually have excellent gluing properties, but they are quite susceptible to furniture beetle attack and the presence of animal glues only increases that susceptibility.

Step 1 Check the Soundness of the Detached Block and Prepare for Re-Use

If a loose block can be re-used (the preferred restoration option), any residual glue should be removed by scrubbing with hot water. Likewise, the old glue should be removed from the case, first by scraping with a sharp chisel and then scrubbing with hot water.

If there are any signs of furniture beetle infestation, the block and the relevant parts of the case should be treated and allowed to dry.

In the event that the glue-block has deteriorated beyond reasonable use, a new copy should be made up in similar wood. Following best-practice restoration guidelines; no attempt should be made to artificially age a new glue block.

Step 2 Glue and Clamp the Block

Whether animal (scotch) glue is used or a more modern equivalent, a thin, flexible lath is used to clamp the glued block in place. Since glue blocks are always on the inside of a box-like structure, it is usually possible to arrange a lath from the diagonally opposite corner. The lath should be flexible enough that it may be bent

Old glue is removed first and the fit checked before re-gluing. Birdmouth clamping blocks are used on the outside of the corner.

into place to exert a diagonal pressure, forcing the glue block into the angle of the joint.

If there is any doubt about the strength of the butt joint between the sheets that the glue block is intended to reinforce, that joint may be clamped as a precaution, prior to inserting the lath.

LOOSE, MISSING OR DAMAGED MOULDINGS

Wood mouldings are used in several locations on the exterior of clock-cases. The base and trunk mouldings are both decorative and functional – they reinforce the load-bearing structure of the case.

On the hood, the applied mouldings are more decorative than functional, with the exception of the skirt around the front and sides of the

To replace the missing trunk moulding, first find a suitably matching piece of wood. Copy the profile from a remaining moulding to a scale-pattern drawing.

Check the pattern against the clock-case.

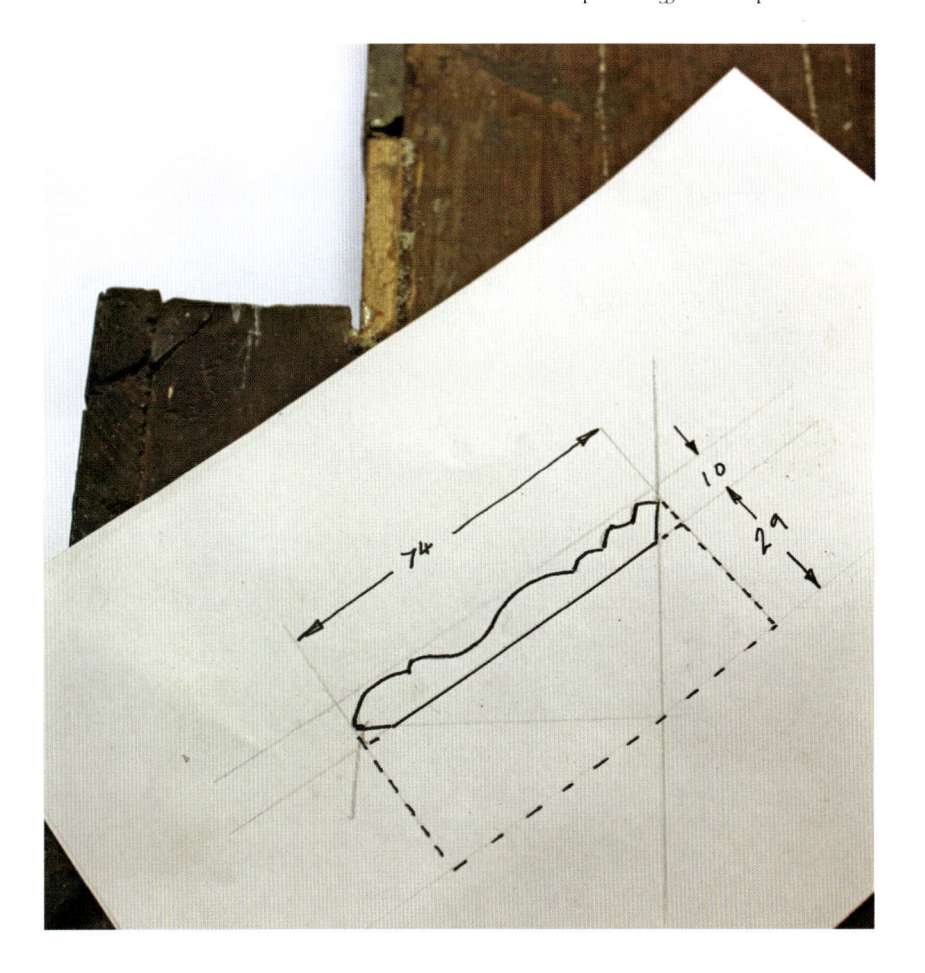

The blank is prepared by gluing the mahogany show wood onto a softwood backing. Early clock-cases rarely used solid wood for mouldings.

With sharp tools, cut the profile, working progressively across the width.

The replacement moulding glued in place should blend exactly at the mitre with the original piece.

A common moulding profile found in all ages of clock-cases. It is simple to make up using a plough plane and sandpaper on the edge of a sheet of suitable wood (8mm thick mahogany in this case). When the profile is formed and finished, it is cut off in a 4mm deep strip.

hood base-plate, which forms an effective dust seal. Applied mouldings were originally glued and nailed to the carcase of the case or hood, and consequently they are apt to loosen or completely detach with long exposure to damp.

Projecting corners of mouldings are very susceptible to impact damage, which usually results in the loss of chips or slivers at the corners.

The neo-classical swan-neck mouldings that were popular, especially for break-arch dial clocks, are particularly susceptible to damage when they are cut with a short grain across the neck of the swan-neck.

Re-Attaching (Re-Gluing) Loose Mouldings

When a moulding section has completely detached, it is likely that it did so because the original glue had broken down. It is unlikely that a satisfactory repair can be achieved while the surfaces are coated with residual glues, so complete removal of the old glue will allow the original gluing process to be repeated. All traces of old glue should be removed from wood surfaces and the parts returned, as far as possible, to the condition they were in when originally glued.

Step 1 Removing Old Animal Glues

A domestic steam-cleaning machine can be useful for removing old glue; the hot steam softens the glue enough for it to be scrubbed away with a stiff bristle brush. Alternatively, a domestic iron applied to a damp cloth (old towelling is ideal) will do the same job but in a more localized, controlled way. Scraping old glue with a craft knife blade is not ideal because it leaves glue attached to the pore structure and depressions in the wood surface.

Step 2 Trial Assembly and Clamping

Tight clamping is essential for effective gluing. When the glue is applied to the mating pieces, it is useful to know that the pieces will fit properly and that they can be clamped effectively without distortion. A practice run with the clamps may seem to be bothersome, but is far less trouble than encountering a problem after the glue has been applied.

Step 3 Glue and Clamp

If animal glue is used, it should be hand hot and, once smeared onto the components, they are pressed together and clamped. Excess glue should be removed as soon as the clamps are applied. Animal glues should be left for a few hours; ideally, overnight.

The procedure for synthetic glues is simpler because they are applied cold. Both resin and polyurethane glues allow time for assembly but, as with all glues, tight, effective clamping is essential.

Step 4 Polyurethane Glues – Remove Excess Glue and Make Good

Although polyurethane glues are excellent adhesives, they tend to foam out of the clamped joint onto the wood surface. The effect can be minimized, but not eradicated, by only using the minimum amount of glue. The best

Excess glue is an unfortunate side-effect of using polyurethane glues. Once the glue has thoroughly cured, it can be removed more easily from polished or waxed surfaces.

option is usually to wait a few hours until the glue has thoroughly cured and then pare the excess foamed glue away with a scalpel and fine tweezers.

Excess glue will peel away quite easily from shellac polish.

THE TRUNK AND HOOD DOOR HINGES

Although there are many styles of both trunk and hood door hinges, the most common types of each differ from ordinary butt cabinet door hinges in that the pin is some distance from the edge of the door and the frame into which it closes.

Trunk Door Hinges

The most frequent faults are loose screws and broken hinge leaves. Occasionally, hinges just wear out through constant use over very long periods – either the pins wear through and break, or the brass body wears. In general, replacement is easier for badly damaged door hinges; modern replacement parts are available from horological and brass-work suppliers.

Hood Door Hinges

The geometry of offset hood door hinges is arranged so that the door closes evenly into its opening. The hinge pin is a screw that fixes the hinge to the woodwork of the hood. They are quite thin and, because they are made from hardened brass to resist bending, they are susceptible to cracking. Repair is unlikely to be successful and in the event of breakage, replacement is a safer option, using approximately 1mm-thick hard brass sheet. The outline of the original is scribed onto brass sheet, cut out with a piercing saw and finished with a fine-toothed file, buffed with emery and finally polished.

Offset clock-door hinges allow the door to close into the opening without binding. They work well when tightly fixed but loosened screws tend to lead to more serious problems.

Strap-type hinges are found occasionally; like all brass, they tend to break and are more difficult to replace.

Offset hinges are peculiar to the cases of longcase clocks. The door hinge is arranged so that the arcs of swing coincide with the opening.

The over-engineered straps on the door part of this type of hinge rarely give trouble. Faults and damage to the hinge can usually be traced back to loose screws on the part that screws to the case.

The hood doors of early clocks often have columns incorporated into the sides of the door. Hinges are not used in that style of making; instead, the door is supported on two pins protruding from the framing of the hood into the ends of the right hand column. In order to remove the door for restoration work, it is usually necessary to locate the end of one of the pins to withdraw it and free the door.

Case-Work – Maintenance and Simple but Effective Repairs 157

RIGHT: Early clock-cases that have the columns fixed to the hood door do not use offset hinges. The hinge pins lie approximately on the axis of the column on the left. The upper pin is fixed into the door and the lower pin is mounted in the hood frame.

BELOW: Turning the hood upside down gives access to the lower hinge pin. In this case a steel screw had broken off.

WOODWORM, WET ROT AND DRY ROT

It is not possible to reverse the damage caused by furniture beetle or fungal decay (wet rot and dry rot), but it is possible to halt the process.

Excellent water-based wood preservatives are available, which will protect even infested or partially decayed wood, but only if the wood can be thoroughly soaked in preservative. In order to introduce the wood treatment into the body of the wood it is necessary, first, to remove the impermeable surface finish. Some risk of damage to the wood finish is implied, so a small sample area should be chosen to test the treatment procedure.

Anobium punctatum is better known as woodworm or furniture beetle. Certain woods are more susceptible to infestation and here, one piece of the hood door was made from oak sap-wood, while the rest of the door made from heart-wood remains untouched.

Step 1 Identify and Remove a Sample Area of the Finish

Using the guidance given previously in this chapter, the finish is removed from a small area, usually on the side-plate of the base.

Step 2 Treat the Sample Area to Test for the Reaction of the Wood to Saturation with Aqueous Solutions

After the starting moisture content is recorded, the test area should be thoroughly saturated with wood preservative. The presence of moisture may have a serious impact on any animal glue that has been used in the construction (including glue used to fix veneers). Drying the wood is a slow process, ideally in a well-ventilated space with no direct heat source. The behaviour of the wood should be checked periodically, while the wood returns to its initial moisture content.

Step 3 Identify and Correct the Side-Effects of Treatment

Although wood preservatives are obtainable as colourless solutions, they are nevertheless

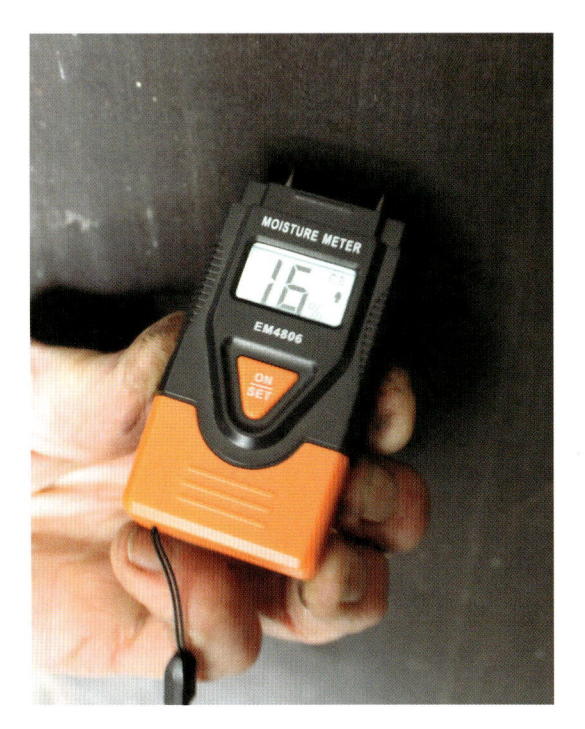

A pocket meter is useful for monitoring changes in wood humidity.

water-based solutions and it is the saturation in water that is likely to cause problems, particularly with the effects of water on animal glues.

Butt and lap joints between wood sheets, glue blocks and veneers should be checked once the wood treatment is dry, in case any remedial re-gluing is required.

Step 4 Introduce Any Work Necessary to Stabilize or Reinforce the Structure

If the structural strength of the case is in doubt, it may be necessary to add new elements or, in extreme cases, an inner frame, in order to stabilize and preserve the outer appearance of the case.

Unfortunately, a great many old and historically significant clock-cases are discarded, but it is quite acceptable to reinforce a damaged case. Quite often some very simple reinforcing will be sufficient to extend the life of an otherwise dangerously unstable clock-case.

Step 5 Check that the Wood Finish Can be Restored Successfully to the Test Area

Having previously established the nature of the wood finish, the same finish is applied to the test area to check that it can be re-finished to match the original.

Old shellac polish tends to become slightly opaque and take on a yellowish grey tone, which will not be present when a new polish is applied. No attempt should be made to colour the new polish to match the old degraded polish – it is quite acceptable that shellac polish requires replacement from time to time.

Wax polishes are a different matter and the loss of patina that has built up over a great many years of rubbing and polishing should be considered. Because wood treatment works best on surfaces from which all waxes and polishes have been removed, the patina will be lost if the wood surface is exposed for treatment.

LOOSE DAMAGED AND MISSING VENEERS

Working with veneers takes a little practice and matching strongly figured veneers, such as burr walnut or mahogany curls, can be achieved with care and perseverance.

A Few Words of Warning About Marquetry

The craft of marquetry, making patterns and pictures using several varieties of wood veneers, was introduced into the English clock trade from the Netherlands in the late seventeenth century. Because it was slow, expensive work, which persisted only until the second decade of the eighteenth century, it is largely associated with good-quality London-made longcase clocks. Consequently, the presence of large areas of intricate marquetry on a clock-case should be taken as an indicator of a high status, high value and an important horological artefact worthy of specialist conservation. However, from the second-half of the eighteenth century, well into the Regency period, it became fashionable to incorporate marquetry motifs into clock-cases: shells, classical urns and cornucopia, sunbursts and fans being the most frequently used. Such inlaid designs should not be confused with earlier marquetry.

Walnut, including bur walnut, was often used to embellished clock-cases before more exotic woods became available.

Bubbled and Flaking Veneers

Changes in moisture content and prolonged exposure to dampness can cause the animal glues to lose adhesion. Coupled with that, the effect of dampness is across-grain expansion of the veneer, resulting in a bubble. When the glue breaks down near the edge of a panel, a flake results that is liable to snag and tear during even the gentlest cleaning.

Bubbles can often be re-fixed quite easily by reheating the glue and applying some pressure. Flakes may be treated in the same way if they are fresh, but when left for long periods, the gradual accumulation of dust under the flake of veneer inhibits satisfactory re-mobilization of the original glue and fresh glue must be introduced.

An ordinary domestic iron provides sufficient heat not only to remobilize old glues, but also to remobilize freshly placed thermo-softening glues or glue film.

Step 1 Iron Back the Bubbles
Protect the surface in the vicinity of the bubble with brown paper or, better, a piece of the brown sheet that is supplied as the backing to glue film.

Set the iron to moderately hot and then press the tip firmly with a circular motion over the area of the bubble.

Remove the brown paper and press the bubble area down with the fingers until the glue has cooled sufficiently to fix the veneer to the ground.

Step 2 Iron Back the Flakes at the Edge of a Panel
Using a paint knife, probe under the flake to find the extent of the cavity. If possible, wrap thin cotton fabric over the paint knife, damp in water and insert into the cavity to remove as much dust as possible. Using the paint knife again, spread thermo-softening glue under the veneer, pressing the veneer down occasionally with the fingers to distribute the glue to the extents of the cavity. Set the piece aside for a few hours or overnight so that the glue dries. (There is no need to clamp the veneer.) Once the glue has dried, the iron is used as before, working towards the edge, so as not to create air bubbles.

Step 3 Check the Effectiveness of the Work
Let the glue dry overnight and then test the adhesion by tapping lightly all over the repaired area with the fingernails or the side of a paint knife. A hollow sound indicates that the veneer is not fixed soundly and further work is required.

Step 4 Further Remedial Work for Stubborn Bubbles
If the ironing process fails a second time, a more serious intervention is required and the veneer over the bubble must be opened in order to insert glue. Carefully slit the veneer bubble and pry up one side enough to slip a paint knife under. Using the paint knife, work some resin-based glue into the cavity and then set the work aside for the glue to dry. Finally, proceed with ironing the mend to re-mobilize the resin glue (as step 2 *above*).

Replacing Lost Veneers

Cross-Bands The practical reason for a cross-band is so that the grain of the veneer meets the edge more or less at a right angle, so it should be more stable and not inclined to split off. Unfortunately, when they become loose, cross-bands tend to detach in sections and the loss of even small pieces from a clock-case can have a serious impact on the overall appearance. Matching cross-band veneers can be problematic – a badly matched mend can be as unsightly as the original fault. A veneer has various properties, each of which should be matched to the original.

Wood Type Cross-band woods tend to have a strong grain structure. Mahogany was used widely on oak cases from the mid-eighteenth century, otherwise walnut is fairly common, satinwood from the 1770s, various exotics and especially rosewood on later clocks.

A figured cross-band on a canted trunk corner is characteristic of certain regional styles. This Irish domestic regulator case has been left in a damp atmosphere long enough for the glued veneers to lift.

Cutting board, straight edge and sharp knife are all that are required. Old veneers that are brittle and prone to splitting can be stabilized before cutting by coating the back surface with the thermo-softening glue that will be used to fix it to the work.

Figure Figure in woods is related to, but not the same as, grain: it is the pattern of colours and textures in the wood that make up its appearance.

Thickness Early veneers were straight-cut using conventional saws, so they may be up to 2mm in thickness. Trimming the thickness of a patch down to the original veneer is not difficult once the patch is in place, but building up the thickness by making up a composite or laminate is more awkward and requires careful measurement.

Orientation The optical properties of certain woods cause variations in the appearance of light and dark areas of the surface. The effect is related to the orientation of the grain or, rather, fibre structures in the wood when the veneer is cut. The results of mis-matching orientation can be quite startling: from one viewpoint, the patch may appear dark against a light background and from a slightly different viewpoint, the opposite. Unfortunately, a wrong orientation only really shows when the patch has been polished and the only remedy is to remove the patch and reverse the orientation.

Lost Lines and Inlays Decorative inlays in the form of lines or bands became fashionable in the Regency period and are often seen in later clock-cases. Although replacement inlays are obtainable from specialist craft suppliers, the range offered is naturally limited. In the event that no suitable replacement is available, sections of inlay may be made up as composite sheets from various components. The work of constructing a replacement inlay is not technically difficult, just slow and exacting work.

Composite bands can be made up by gluing sheets of different wood and then off-cutting slices. Holly and bog-oak are used here to make up black/white and white/black/white line.

Chapter 9

More Difficult Woodwork Procedures

The dilemma about whether to follow the principles of restoration (restoring to some former state) or conservation (conserving what is present) is a feature of planning the maintenance work on the cases of longcase clocks because they are, to varying degrees, historical horological artefacts. Any clock has to be taken on its own merits and there can be no black-and-white answers. Furthermore, the question is not a static one, and with the passing of time, as clocks become increasingly rare, conservation for future generations will increasingly become the preferred choice. The over-riding objective is usually to prevent further deterioration when the structure of a case has become unstable; after that, once a case is stable, questions about restoration or conservation are less pressing.

While the professional conserver has an ethical obligation to at least make his client aware of the conservation/restoration choices in the fundamental approach to repairing a clock-case, for the amateur clock-owner there are no restrictions on the quality and nature of conservation, restoration or repair work, although he or she may hope that his/her clock may, one day, be a source of enrichment to future generations.

Consequently, with all case-work, it is prudent to follow the conservation route as a general rule, and it is useful for the repairer to ask himself from time to time, 'What will my great-grandchildren think of this repair when they become clock connoisseurs?'

SOME BASIC RULES ABOUT CONSERVING AS MUCH AS POSSIBLE OF THE ORIGINAL

Other than a brief appendix on the subject of case restoration, the British Horological Institute publication, *Conservation of Clocks and Watches*, deals principally with clock movements, dials and so on. For guidance that is more specific to furniture, but very relevant to clock-cases, *The Conservation of Furniture* by Nick Umney and Shayne Rivers gives complete and formal, even theoretical, documentary guidance to what is admittedly hands-on work. It sets out the ethical principles of conservation, some of which are:

- The emphasis on preservation for the future, accepting the demands of use.
- Understanding and appreciation.
- An emphasis on preventive conservation.
- Documentation of the nature and extent of conservation work.
- The principle that work need not be obtrusive but should be detectable.
- The keeping of intervention to a minimum.
- The retention of as much original material as possible and the addition of as little new material as possible.
- The avoidance of irreversible alterations.

HOW TO DECIDE WHETHER OR NOT STRUCTURAL WORK IS REQUIRED

A thorough assessment of the degree of instability is fundamental when considering mending a clock-case. Stability in this context means the structural stability necessary to support the weight of the clock with its driving weights without movement, distortion or even complete collapse, and also the stability of the surfaces, veneers and other non-structural embellishments.

Early cases tend to have long, wide, door openings with narrow facings to the front of the trunk. Consequently they are intrinsically less rigid than later cases. A well-fixed backboard is critical to the overall stability of the case and its resistance to twisting. (Steve Daniels)

A clock-case may usually be considered unstable if:

- There is any noticeable static deformation; for example, the trunk door no longer closes properly into its opening or the base is no longer true with the trunk.
- There is any tendency for the case to twist under hand-pressure when the base is held firmly and, with the hands on the sides of the trunk, a rotational force is applied. The

This early eighteenth-century case is quite battered but structurally sound. The untidy appearance could be improved by some cleaning, but nothing more needs to be done.

Testing the glued joints with a paint knife will quickly show up incipient problems.

flexing of the case, which will be seen usually at the joint between trunk and base, may indicate that the case is no longer a rigid three-dimensional structure, as originally made. Gaps in the mitred joints of the trunk mouldings usually imply that sort of twisting deformation.
- There are gaps into which a paint-knife or similar thin blade may be inserted; for example, between trunk and trunk mouldings.

PROPER ASSESSMENT

Before the computer age, restorers often made detailed sketches and notes, primarily to list every fault for pricing and planning the proposed restoration work and, second, as a matter of routine record-keeping.

Nowadays recording is made much easier with annotated digital photography. Whichever recording method the restorer uses, the procedure of identifying, illustrating and describing the faults and previous repairs is likely to result in a full description (what restorers and conservers call a Condition Report) of the clock-case. Without such a rigorous approach, it is very easy to overlook apparently unrelated faults.

The professional restorer is obliged to listen to the client's ideas about what is wrong with the clock-case and how much the budget will run to. In many cases, the work must be prioritized in favour of stability, even at the (temporary) expense of appearance. By comparison, the amateur owner/restorer is often apt to put effort into enhancing the appearance without fully appreciating the significance of structural faults.

So, for the amateur restorer, the preparation of a condition report not only records for future generations what is there at present, but forces a detailed all-over inspection, which, with the passing of years, will become a valuable document, if it is kept with the clock.

PARTIAL OR FULL REPLACEMENT OF THE BACKBOARD

The backboard closes the dust-free containment of the clock-case and, as a structural element, it fixes the two side-plates of the trunk and unites the trunk with the base, transmitting the weight of the clock from seat-board to floor. Inferior woods were often used to make up the backboard – various softwoods or sapwood, which, apart from being mechanically weak and prone to splitting, are vulnerable to rot and woodworm infestation. In country clocks, the backboard often extended to form the back feet of the case, and the resulting vertical grain alignment contributed to the rise of moisture up the case.

A clock-case is supposed to protect its clock from dust. The losses in this backboard have contributed to wear in the clock movement. Immediate repair is required.

In most cases of failure, the lower section of the baseboard is most likely to have deteriorated beyond the point of usefulness. A further complication in that section of the baseboard is caused by the way that the backboard is widened across the back of the base.

Consequently, when judgements are required about how much to keep and how much to replace, the deciding factor is often the location

More Difficult Woodwork Procedures 165

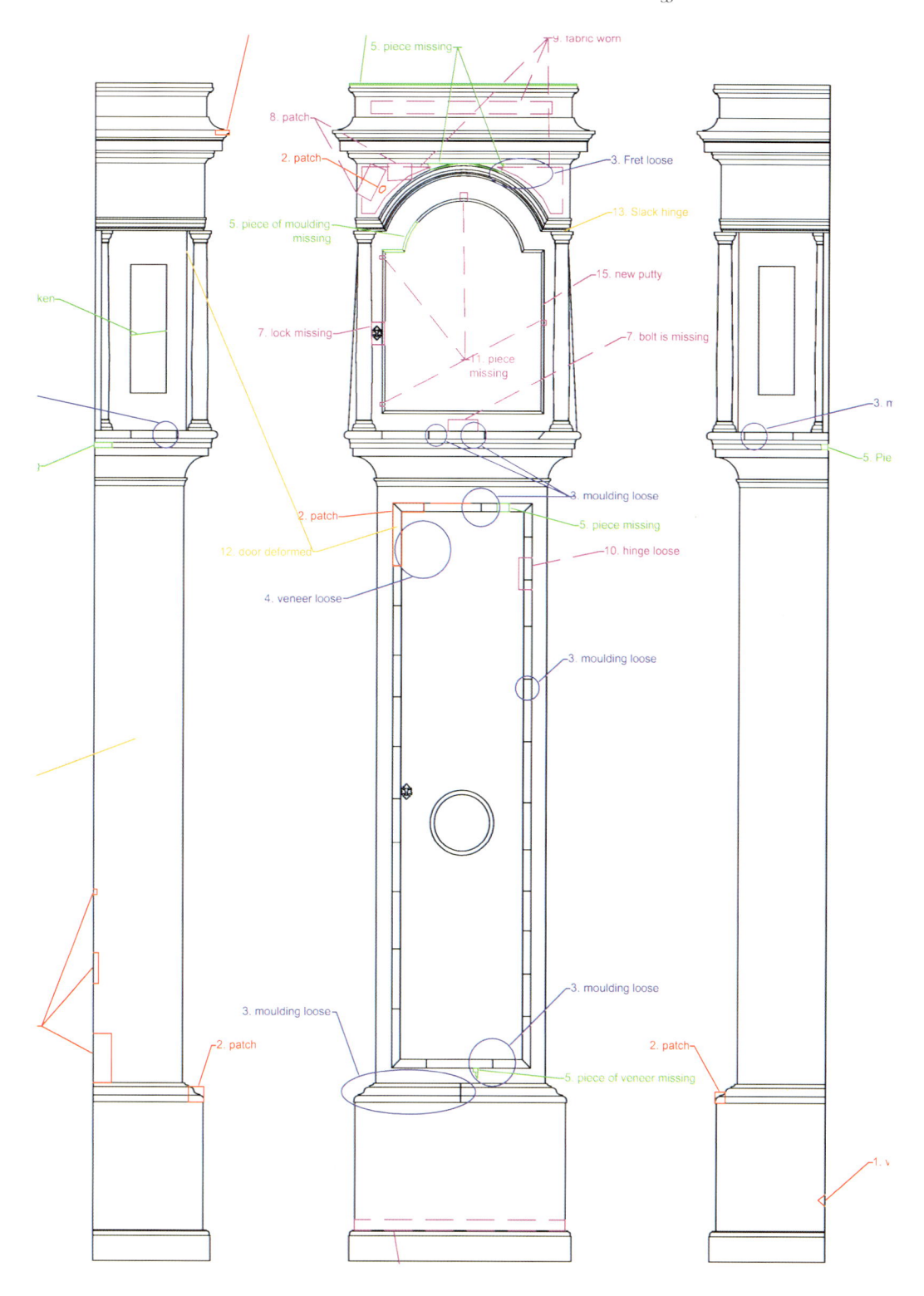

A conservation project calls for detailed annotated drawings, photographs and written descriptions. Time and cost are implied, and the budget should reflect the paperwork.

A common cause of problems with the backboard. The fill-in pieces, where the base section widens, are often loose or missing. In this late eighteenth-century case, the split and loose boards are re-glued.

of the lowest sound piece of the backboard. Once that position has been decided on, a design for the joint between old and new work will be required, so that the old and new sides of the proposed joint may be cut in preparation.

The objective of the work being long-term stability of the case, the considerations in selecting the most appropriate wood to use for the replacement part are:

- **The wood type** – matching oak with oak or deal with deal. Aside from minimizing obtrusiveness and keeping to the original materials, like-for-like replacement will minimize the risk of future differential movement between the new and original parts.
- **The nature of the wood** – it is tempting to use old wood, perhaps even of the age of the clock-case, but some wood, especially the softwoods, often lose natural spring or 'nature', becoming brittle and biscuit-like in texture. When a sample of the replacement wood is cut and shaped it will usually be obvious if the wood has deteriorated not only by the absence of the slight resinous smell of good quality deals or the characteristic tannin smell of oak, but also by the way the wood can be cut with a sharp blade.

Step 1 Plan and Prepare the Work

The rigorous approach of the professional conserver requiring detailed documented assessment and a well thought-out action plan is likely to result in successful restoration work. Preparing a few detailed drawings of how the replacement part will fit into the case will often bring to light questions about the construction detail, such as how to clamp up the work. It will often be necessary to glue up the work in a number of stages: glue blocks are fixed to the carcase first and the replacement section of backboard secured later, once the first gluing is properly set.

More Difficult Woodwork Procedures 167

During the process of removal of the damaged parts and preparation of any jointing surfaces, it is useful to record details onto the design sketches, which should be preserved with the clock. Where the side-boards of the case have been rebated to take the backboard, care should be taken to remove any old fixing nails that will prevent the replacement part from seating correctly.

Step 2 Make Up the Parts and Assemble
Accurate, dimensional sketches will greatly increase the efficiency of cutting and preparing the replacement parts. The detail of the joint between old and new work will have been thought out and, where a staff joint is used, the angle will be set on a bevel gauge.

Step 3 Trim and Make Good
Careful planning and accurate work will minimize the amount of work in cleaning up the repair. Lap and staff joints, where the two gluing surfaces are prepared with inclined cuts, tend to bleed surplus glue, which should be removed because it is unsightly and obtrusive. The repair should be finished in a manner that harmonizes with the original, so colouring may be required. The conservation rule that calls for unobtrusive, without any attempt to disguise, is difficult to apply in practice, especially for the amateur clock-owner, but repair work that is neat and well-executed does not necessarily detract from the value of a clock-case.

REBUILD THE BASE

Because the base is more prone to mechanical damage (collisions, etc.) and rot from the prolonged effects of damp floors, the base structure of a clock-case is prone to loosening of glue blocks and the glued joints between framing, sheets (the outer surfaces) and mouldings and so on, and from woodworm infestation.

When the damage is more widespread than simple isolated areas, the easiest and most effective procedure is to completely rebuild from the trunk and backboard to the original design, using as much as possible of the original components. It is not unusual for the base to

A new piece is let in to the backboard to close it and give lateral support and improve rigidity.

The repaired backboard (see page 164) is now a reasonable fit with the hood. The objective of colouring new wood is not to pass it off as old – just to make it less obtrusive.

disappear completely; the detached components of severely damaged cases tend to get separated and the parts lost.

Step 1 Record and Dismantle the Remnants of the Old Base

Accurate drawings are the starting point for designing replacement parts and the whole case should be inspected thoroughly to ensure that it is completely sound.

Any remnants of the old base, including parts of mouldings, should be saved as patterns for the new work.

Step 2 Plan the Work and Materials for the New Replacement

The case itself will contain features that are copied in the base; the overall style will suggest whether or not feet or a plinth were most likely. Square corners to the trunk are usually accompanied by similar corners in the base but more elaborate trunks, especially in later

A loose cross-band has been removed to reveal a serious woodworm infestation. The base is structural and must be capable of supporting the clock, including the driving weights. Interesting conservation questions.

More Difficult Woodwork Procedures 169

The base of this third-period clock has completely disappeared. A replacement must be designed to reflect and harmonize with the style of the remaining parts.

Two frames are used in the base assembly. The upper one is open to allow for the fall of the weights.

northern clocks, usually have canted corners to the trunk.

The planning and design stage should result in sketches that show how the complete case most likely looked in its original condition, so it is useful to have access to as many reference books as possible.

To achieve a good match in appearance between the original parts of the case and the replacement base, it will be necessary to locate the right woods, including the inlays.

Patterns for details such as feet and moulding profiles should be taken from cases of similar region, age and status.

The upper frame comprises three sides and is made up to suit the outside of the trunk carcase, to which it must be securely fixed. The detail of any canting, the thickness of the sheets that make up the base, the trunk moulding and the degree to which it overhangs the base-sheets, are all design considerations.

The lower frame usually has three sides but occasionally is secured to the backboard by a fourth. In plan view it coincides exactly with the upper base-frame and a solid base-board adds rigidity and prevents dust getting into the case.

Step 3 Make Up and Fit Framing

A good-quality, close-grained deal is ideal for the framing; it works accurately with clean, sharp edges and takes glue well. The parts of the upper frame should be fitted individually to the trunk, taking care that the trunk is square.

The lower frame can be made up prior to fitting to the backboard.

Step 4 Make Up and Assemble Front- and Side-Sheets

The side- and front-sheets of the base may include veneers, inlays or cross-bands. These decorative details are most easily applied on the bench before the sheet is attached to the case.

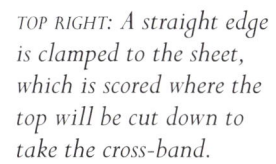

TOP RIGHT: *A straight edge is clamped to the sheet, which is scored where the top will be cut down to take the cross-band.*

ABOVE LEFT: *A V-cut is made to save the edge of the sheet.*

RIGHT: *A rebate plane is used to remove the thickness of the cross-band.*

172   *More Difficult Woodwork Procedures*

LEFT: *A straight edge is formed along the edge of the new cross-band.*

BELOW LEFT: *The cut should be clean and even.*

BELOW RIGHT: *The new cross-band is left wide and trimmed back after gluing.*

A typical cross-band, such as a 1in-wide mahogany band around the edge of the front-sheet, is applied by first cutting the surface with a cutting gauge or marking knife and then cutting down the surface to the depth of the band using a rebate plane.

The cross-band, which may be new veneer or salvage from well-figured veneers on scrap furniture, is glued in to the cut surface with mitred joints at the corners.

In mahogany veneered cases, the front of the base may contain a panel that matches the trunk door, often using a very decorative flame veneer. As a general rule, the most strongly figured veneers are the most difficult to match but, with patience, it is usually possible to find a good match.

Step 5 Assemble Base Carcase
Once the decoration of the side-sheets is complete, the base may be assembled. It may be preferable for alignment and clamping to fix the front-sheet first and, once that glue is set, fit the side-sheets later.

It is natural for distortion to have developed in the original trunk and, consequently, some adjustment may be necessary when the sheets are fitted: mitred joints on sheets are often problematic and a stepped joint at the corners is easier and neater with the joint line a few millimetres back from the front edge of the side-sheet.

Step 6 Fit Trunk Mouldings
The trunk moulding reinforces the joint between the base and the trunk, so it is structural as well as decorative. The mitred joints should be tight and, once the moulding is fixed, the sharp edge at the mitre is softened with fine sandpaper and re-coloured, as necessary.

Step 7 Finishes
Finishing new work to match is always difficult and a noticeable lack of harmony between the new and the original is often alleviated by re-polishing the original.

REBUILD THE TRUNK

Purposeful dismantling is a very last resort but, occasionally, some catastrophe results in a case becoming very badly damaged, even smashed to pieces. Falling over generally, falling down the stairs, falling during road transport and even dismantling as part of an abandoned amateur restoration project, can all result in a box-full of disconnected, damaged parts. Rebuilding may replicate the original making but the presence of marquetry, inlays and decorative veneers may complicate matters because, in the original making, they were applied to the structure of the case after it was assembled, often crossing construction joints.

Step 1 Lay Out the Parts and Identify the Losses
Even quite insignificant parts can be very useful: glue blocks and fragments of beadings and veneers. Once the parts are laid out, it is easy to see what is missing.

Step 2 Clean Up and Prepare the Backboard
It is important to know that, when the trunk is rebuilt, it will fit exactly onto the backboard, so preparing that part is the easiest way to proceed.

If any new sections are required, they are fitted after the splits or cracks in the boards are glued up. If reinforcing is necessary, thin strips of wood fixed across the grain will hold the backboard together and, from a conservation point of view, a reinforced original backboard may be preferable to a new one. Fixing nails that originally attached the backboard to the trunk are removed or filed flush with the wood.

Step 3 Fix the Trunk Components Together and Attach to the Backboard
The front-plate of the trunk is usually fixed to the sides by a series of glue blocks and, in re-assembling the case, it is much easier to fix the glue blocks to the front-plate first and then attach the sides to the front-plate. With the prepared backboard to hand, it is quite straightforward to clamp across the two side-plates using the backboard, loosely in place, to provide the reaction to the clamps.

The glued joints above the level of the hood moulding between the top of the front- and side-plates should not be overlooked.

Once the glue is dry, the work is checked and, if all is well, the backboard is re-attached to the side-plates.

Step 4 Attach the Base Framing

The re-assembled trunk is quite delicate without the stability given by the two rectangular frames of the base. The upper frame is attached to the outer faces of the bottom ends of the side- and front-plates. The strength of the jointing at the corners is crucial to the long-term stability of the case. New framing is often required but, if the trunk moulding is still attached to the original frame, it should be carefully removed so that the frame may be jointed soundly.

The most successful arrangement for the lower base-frame is a four-piece rectangular frame that unites the base to the backboard and supports the feet or plinth; it takes its strength from the base-sheet. In order to facilitate the process of fixing the front and sides of the base, a continuous glue-block may be fixed between the frames, prior to attaching the panels.

Step 5 Attach the Base-Sheet and Feet/Skirt

The base-sheet is often overlooked but it keeps the case rigid and prevents the ingress of dust. Once it is attached to the lower base-frame, the base will be rigid enough to accept the feet or plinth, depending on the style (as an approximate rule, southern cases tend to have a plinth or skirting, while feet are more likely in northern cases).

Step 6 Attach Trunk and Hood Mouldings

The hood moulding supports the weight of the hood and contributes to the overall rigidity of the case; likewise, although the trunk moulding masks the joint between the base and the trunk it reinforces the base-framing, adding to the overall stability of the case. Consequently, the closeness of fit of the mitred joints will affect both the appearance and the structural strength of the case.

A continuous glue block fixed between the upper and lower frames keeps the whole sub-structure rigid.

Step 7 Patch Veneers and Inlays

Patching and matching veneers is never easy because every piece of every tree is different, but with perseverance, and a stock of old furniture and scrap veneers, it is usually possible to match veneers reasonably inconspicuously. Colour matching is often a problem, especially when grime and old shellac polish masks the true colour of the veneer. As it ages, shellac polish tends to develop a yellowish, sometimes grey tint and it often becomes opaque. Rubbing back with spirit (ordinary alcohol) on a cotton rag or very fine wire wool will dissolve and remove the old polish.

Inlaid decorations in clock-cases are extremely difficult to repair but, again, with a lot of practice, and a wide choice of veneers, good matches are possible.

Before attempting to cut small pieces of veneer for patching, it is useful to paint a thermo-softening glue onto the back. Once the glue dries, it will hold the veneer together and, when the patch is inserted into the similarly treated repair area, it is fixed by pressing with a hot iron, which re-mobilizes the glue.

Inlaid decorations are made up from featureless woods, e.g. box, pear wood and sycamore. They are coloured to match and the shaded effect is achieved by charring the edge in hot sand to darken the wood.

Mahogany flame veneer. The panel in the base matches that in the door. The most likely source of veneers of that figure is in scrap antique furniture, such as Victorian mahogany wardrobes.

This hood has lost the upper and lower sections of its bezel. Without the rigidity given by the bezel, the hood is extremely fragile.

Step 8 Finishes

The finish that is applied should match the overall style of the clock. Mahogany and other exotics should have bright, reflective shellac finishes, while oak cases, especially those of country clocks, are usually better waxed.

REBUILD THE HOOD

Hoods often become seriously damaged and, unlike the case, which is a fairly rigid box structure, the hood is more or less open on three sides. It is intended to be light enough to be removed easily from the case and, consequently, the hood can be quite flimsy, relying on clever design for structural rigidity. A poorly appreciated component is the bezel, a thin sheet that fills the space between the sides of the hood and the dial. Without the bezel to hold the dial opening square, a square dial or break-arch hood will tend to deform with the door open. The hoods of round-dial clocks tend to be intrinsically more stable.

The structural base of the hood is the three-sided frame that rests on the hood moulding of the case. Case-makers of different abilities and conditions traditionally used a variety of joints, ranging from the country carpenter's simple halved, angle joints that rely on a glue bond, to the cabinet-maker's mitred bridle joints. Likewise, the side-sheets are either mortised into the base-frame or, in more vernacular clocks, nailed to the inside face with a separate runner. In either case, the hood is arranged to slide forwards off the case without tipping forwards. The pediment provides additional rigidity, provided that it is fixed tightly to the side-sheets.

Round hoods, often termed 'drumhead', rely on a pair of hoop-shaped wood frames and a solid front-sheet or bezel for rigidity. The body of the round hood is covered over with coopered strips of wood, which, when fixed tightly to the hoop frames, ensure a strong dust-proof top.

A typical early feature – a view from underneath shows the mortise joints that hold the side-sheets to the base-frame of the hood.

A replacement hood made up for a drum-head clock. The structure of the hood relies on two wood hoops onto which the coopered panels and the bezel are glued. The door is shown fully open and the decorative sill and lower mouldings have yet to be fixed.

Step 1 Check the Fit of the Base-Frame in the Case

A badly fitting hood is likely to become damaged by rough handling during removal, and over the years distortions and repairs to the case often result in a lack of square in the moulding that supports the hood. The base-frame should be an easy, sliding fit, restrained above by runners fitted to the upstanding side-plates of the case. If the base-frame does not slide fully and easily into the closed position, it should be examined very carefully and adjusted if necessary.

Hood mouldings tend to get re-fixed over the years and while checking the fit of the base-frame it is worthwhile to check that the moulding is at right angles to the trunk, both sides and the front.

Step 2 Attach the Side-Sheets to the Base-Frame

The side-sheets that form the enclosure of the hood are fixed rigidly to the base-frame. When they are nailed to the inner faces of the base-frame, a separate runner is fixed inside. Otherwise, on better made cases, the side-sheets are mortised into the base-frame. Prior to re-assembly, old glue should be removed and, if the original tenons are broken, new ones are dove-tailed into the side-sheets.

Step 3 Re-Assemble the Bezel and Pediment

The lower section of the bezel is first glued to the front section of the base-frame. Then, in order to avoid distortions as the hood is re-assembled, a temporary sheet of plywood is cut to exactly fit between the side-sheets; it is clamped to the back of the base-frame to ensure the that the side-sheets are vertical.

The pediment is supported at each side by soffit pieces fixed to the tops of the side-sheets and rigidity is achieved by the completion of the bezel.

Step 4 Attach the Dust-Proof Top

The practical purpose of the clock-case is to prevent airborne dust from settling on the clock movement but, additionally, as a structural component, the top of the hood gives the whole hood rigidity. So if the hood top is loose, the rigidity is jeopardized and, of course, dust will enter the clock movement.

Occasionally, a piece of fabric is found glued to the top of the hood, fixing the individual pieces together. Whether original or not, it is a highly effective way of stiffening and dust-proofing the hood.

Step 5 Fit the Hood Door

Most, but not all, hood doors are fitted with off-set hinge plates that allow the door to swing forwards clear of the hood structure. Round-dial cases hinge either from the side (always the right-hand side) using a pair of small brass butt

The same principles apply to plain regulator-type hoods. A bezel behind the door keeps the hood rigid and the door is fixed by small brass butt hinges.

hinges or, in the case of drum-head type cases, from the top using a single butt hinge.

With the hinges fitted to the door, it is placed into the opening of the hood with card spacers around the sides, the hinges are screwed to the hood and the clearance checked. The hinge screws should be a good fit to the holes in the hinge plates or the door will be apt to drag on the bottom of the opening.

Step 6 Restoring and Making the Accessories

COLUMNS (INCLUDING BARLEY-SUGAR TWIST), MOULDINGS, SWAN-NECKS, PATERAE AND FINIALS

Turnery is an old-fashioned word for any object turned on a lathe. In the cases of longcase clocks, the turned objects are columns either side of the hood door, free-standing columns, half-columns or quarter-columns either side of the trunk door. Single barley-sugar twist, often just called single barley twist or even single twist, is often found on older cases and was quite a common feature prior to the adoption of neo-classical styles in the second-half of the eighteenth century.

Making up a replacement single twist is slow work but, with practice, exact copies are possible.

Making Up Single Twist
Step 1 Make Up the Blank
It is essential that when the twist is being cut in the lathe, the blank has a perfectly square, roughly 55mm extension at each end, so that the blank can be turned by hand and twisted against the cutting tool.

Step 2 Turn Down the Diameter of the Twist and Cut in the Coves
The blank is mounted between centres and reduced to the outside diameter of the twist. At each end, the cove is cut down to the inner or lesser diameter, defining the length of the twist. The hollow should merge with the cove, so cutting the cove first gives a good indication of the desired depth of the hollow.

Step 3 Mark Axis and Pitch Lines and Draw in the Centres of the Bine and Hollow
Using the tool rest as a straight edge, the cylinder is marked with eight axis lines and then pitch lines at one-eighth of the pitch of the twist, resulting in a pattern of rectangles over the surface of the cylinder.

Twists usually come in left- and right-hand pairs, so first decide which twist and then form the centre-line of the hollow by joining diagonal intersections. Draw a similar line four pitch lines along, which will define the top of the bine (the bead or raised part that spirals around and runs the full length of a twist).

Setting out the work is the key to precise woodwork. Cut away the cove but leave the square ends on the blank to facilitate marking the eight axis lines.

Making twists takes practice. The tools are short and are just held against the tool rest of the lathe with one hand, while turning the work onto the bale with the other. There is no need to push with the cutting tools, and reversing the piece frequently will give even progress.

Step 4 Make a Saw-Cut Along the Line of the Hollow and Rough Out the Twist

Gripping the square end of the blank with the left hand (for right-handed people), make an initial saw-cut to an even depth along the marked cut-line, using a saw with a wide-tooth set. The saw-cut should extend to about two-thirds of the depth of the hollow. It is possible to make or adapt a saw to cut to a limited depth, but masking tape on the blade works.

Using a gouge, and cutting 'down the grain', a margin of a few millimetres is removed from one side of the saw-cut by twisting the blank with the left hand into the cutting edge. When the full length of the blank is cut, it is turned end-for-end in the lathe and the first gouge cut repeated on the opposite side of the saw-cut.

A twisting gouge, which has a very short blade, is much more controllable and easier to use than a normal turning gouge but in either case, the tool should be very sharp.

The third cut of the gouge is through the bottom of the hollow to the depth of the saw-cut.

Step 5 Finishing the Twist

Continue to take cuts from each side and the bottom of the hollow, leaving an uncut band at least a quarter of the pitch wide, centred on the centre-line of the bine.

Once the depth of the hollow is cut, its circular profile is roughed out using a very sharp, wide, straight wood chisel; with the bevel against the cut and the blade held nearly vertical, the chisel will plane off the sides of the cut as the blank is twisted.

The sharp edges of the bine are rounded using a palm plane (a block plane will do), always working 'down the grain'.

Finally, the twist is sanded, starting with a coarse paper and working down to very fine, taking care to maintain the sharp edge where the bine blends into the cove.

The finished twist is coloured to match the original using oxidizing agents (potassium permanganate for oak or potassium dichromate for mahogany) followed by wood stains or dyes, which are obtainable from DIY shops. It is prudent to try the colouring agent on a few samples.

Seal the grain with shellac and then finish by polishing, either with shellac or beeswax, to match the original.

Step 6 Cutting Half-, Quarter- and Three-Quarter Twists

Half- and quarter- and even three-quarter twist columns are often found in clock-case hoods. The square ends are left on the blank to enable it to be cut true.

The easiest way of cutting half- and quarter-columns is in a band-saw and a straight, true wood strip is glued to the square ends and used to run against the fence of a band-saw.

The hood columns in older cases often are incorporated into the hood door as three-quarter columns and making replacement parts requires extreme care. Once the column has been turned, two intersecting saw-cuts are made along the length of the column using a hand-saw. Setting up the workpiece on the

This clock dates from the mid-1770s but the style is older. The twists are in opposite pairs (right and left hand threads together); the door pieces are three-quarters and the back ones are halves.

Finish with rasps followed by coarse and fine sandpaper.

bench is critical so that the saw-cuts can be made accurately. A very sharp, fine-tooth saw should be used and, if several three-quarter columns are required, it may be worth making up a long, shallow tenon-saw especially for the purpose.

DESIGNING, MAKING UP AND FITTING REPLACEMENT PLINTHS AND FEET

Early Dutch longcase clocks of the late seventeenth century often stand on feet but typical London longcase clocks of the first two decades of the eighteenth century are more likely to stand on a plinth. There are many exceptions documented by Brian Loomes, Cesinsky and Webster and others, but it seems clear that, as the eighteenth century progressed, a geographic division appeared with plinths favoured in the south and feet in the northern counties, Scotland and Ireland, so that by the nineteenth century, a plinth on a northern clock would be very unusual.

The plinth is an element in the aesthetic composition but it is both structural and functional: it gives stability and protects the more delicate parts of the case from accidental damage. Consequently, it is a part of the case that is likely to have sustained damage and is also likely to have been repaired or replaced, perhaps on a number of occasions. Country clocks are particularly prone to losses of plinths or feet, either through inferior materials and craftsmanship or as a result of harsher conditions.

Sometimes there are clues in the base that might shed light on the design of a former plinth or feet but, in general, looking at the overall style of the case in the context of contemporary fashions and regional variations will provide sufficient evidence.

The moulding on the left is a type that was used above separate feet, which are now missing. On the right, the front panel of the plinth on this third-period Suffolk clock has been opened out to give the impression of feet.

Hybrid plinths with cut feet are typical of mid- to late eighteenth-century clocks. The designs tend to be based on contemporary furniture; the example on the right resembles the feet of a mid-Georgian chest of drawers.

Once the overall appearance of a replacement plinth has been decided upon, consideration should be given to how it will support the weight of the clock. In some cases, the plinth is merely applied to the lower section of the base panels, which themselves bear the weight. Otherwise it may be necessary to cut a rebate on the inner face to support the weight.

The cut moulding along the top edge of the plinth serves only to break up the heavy appearance, but without that detail a plinth is likely to look awkward or cumbersome.

Replacement feet can be more problematic, both in deciding upon a design and also for a method of fixing securely to the case. Eighteenth-century longcase clock feet tend to be smaller but similar in shape to the feet on contemporary furniture, especially chests of

Shaped and mitred bracket feet were especially popular in later northern clocks.

The wood should be chosen for the way it cuts and finishes. A very basic set of sharp carving tools will be sufficient to form the various elements.

drawers. Front feet are usually made up from two parts, mitred together, while back feet are single pieces. It is not unusual for a clock-case to have just two front feet.

MAKING A REPLACEMENT SWAN-NECK MOULDING

The introduction of the swan-neck pediment in clock-cases followed on from the adoption of neo-classical styles in furniture design and the popularization of the break-arch dial in the third-decade of the eighteenth century. Early break-arch cases tended to retain caddy or pagoda tops but then an aesthetic shift occurred from the earlier Queen Anne/George I styles to the distinctive Georgian neo-classical. Although it was occasionally used in square-dial cases, the swan-neck pediment became the norm for break-arch cases, particularly in country clocks. Consequently, because 30-hour clocks tend (with many exceptions) to have square dials, the swan-neck pediment tends to be associated with eight-day movements.

From simple cut-out softwood shapes in

country clock-cases to intricately carved mahogany, the swan-neck pediment is intrinsically prone to splitting along the short grain. With prompt action following a mishap, sound, unnoticeable glue repairs are usually possible, but unfortunately the broken off piece is often lost.

Step 1 Match the Wood Type and Design the Part

It should go without saying that any slight discrepancy in the symmetry of the pediment will be immediately obvious, so great care should be taken in setting the design of a replacement part. Equally, the colour and overall appearance of the replacement should match exactly, so a comparable wood is essential. The veneer on flat swan-necks can be difficult to match.

Once a paper pattern is made, it is transferred to the wood blank.

Step 2 Cutting and Shaping

The blank is carefully cut out using a band-saw or otherwise a coping-saw. The lower end is mitred onto a top moulding on the side of the hood, consequently that moulding should be used as the guide. The salient points of the profile are marked out in pencil and the detail is cut with a few very sharp carving tools. The circular termination at the top end of the pediment is the part most likely to be lost and, in keeping with the concept of preventative conservation, as a precaution against future loss or damage, it may be prudent to glue a thin sheet onto the back of the pediment with the graining across that of the pediment.

Step 3 Finishing

Once the form of the swan-neck is formed, it should be sanded to a smooth finish, which will show up any imperfections. A good colour match is usually easier with mahogany. It is first treated with potassium dichromate and, when dry, a rub of linseed oil will give a good indica-

The replacement part pictured on page 184 coloured and finished and fixed to this third-period clock. The brass-ware is new.

tion of the final colour. Wood dyes are suitable for the slight colour adjustments necessary to bring the new work to an unobtrusive match.

Step 4 Embellishments

The circular termination of the swan neck is invariably embellished in some way – the only exceptions are country cases made up in softwoods. Pressed brass paterae were the most widely used, and modern reproductions are available. Otherwise, gilded turned bosses or carved flowers were used, all of which are, at least theoretically, reproducible.

Glossary

Addendum
In gear teeth, the distance that the tips of the teeth are above the pitch circle.

Anchor escapement
First patented in 1666 by Robert Hooke, the anchor escapement allows for small pendulum swings and is ideally matched with long pendulums. The usual type of anchor escapement comprises two pallets set at an angle of 90 degrees, that span one-quarter of the escape wheel (typically, but not always, 7½ teeth for a 30-tooth wheel) and allow the wheel to advance by a half-tooth as it interacts with the pallets alternately. Its action gives rise to an apparent recoil of the seconds hand, which regresses slightly after each advance of one second, giving a rather imprecise indication.

Arabic numerals
The ordinary numbers that we use every day. In longcase clocks, Roman numerals were used for hours (except for a few years at the beginning of the nineteenth century) and Arabic numerals are used for days of the month, minutes and seconds.

Arbor
Clockmakers term for the spindle or axle on which wheels and pinions are mounted.

Arch dial (also known as break-arch dial)
Early longcase dials were square until the early eighteenth century when arches were added. By about 1720, square dials were rare in London clocks, although they continued in country clocks. The centre of early arches tends to be below the top edge of the square.

Back-cock
A bracket screwed to the outside, top-back of a clock movement. It supports the pendulum in a pair of chops (a rearwards extension) and usually supports the rear escapement arbor pivot.

Bolt and shutter
An early form of maintaining-power device designed to keep power on the going or time train of wheels during winding. An anchor escapement clock tends to run backwards during winding and deadbeat escapement clocks are prone to damage if the impulse to the pendulum is lost. Bolt and shutter is only found on early, good-quality clocks and was superseded by the Harrison maintaining-power device; its presence is suggested by recessed winding squares and a plate (the shutter) behind the winding holes.

Bur (burr)
Bur has two meanings: the rough edge on a piece of metal, including the cutting edge of a cabinet scraper; also, a decorative figure in certain woods, especially walnut.

Bush, bushing (occasionally, bouchon)
Strictly speaking, a bushing is a replacement sleeve or tube of brass inserted in a hole in a movement plate that supports pivots. However, in a more general use, it is the hole itself. The consequence of the re-bushing technique is that clock movement plates tend to be infinitely repairable. Every time a hole becomes worn (typically oval), it is opened and re-centred and the new bush pressed in.

Cannon wheel
Another name for minute wheel, the wheel

mounted on the centre arbor that carries the minute hand on a pipe.

Chapter ring
A brass ring, usually silvered mounted on the dial plate and engraved with hours and minutes.

Chops
An older clockmakers' term for the jaws of a vice, it is used more widely for any facing surfaces designed to support or contain something such as the split pendulum suspension bracket on the back-cock of a longcase clock.

Clock or timepiece
A clock has a bell (the word clock is derived from the Latin word 'clocca', a bell). A timepiece has no bell and only shows the time.

Collet
A flanged sleeve that fits over an arbor to support a wheel. Early clocks relied on silver solder for fixing, but in later clocks the collet is soft soldered to the arbor and the wheel is likewise soft soldered to the collet.

Count-wheel
A notched wheel that controls the strike cycle (i.e. it counts off the strikes at each hour). When a latch or detent drops into a notch, another linked detent in the strike-work is allowed to fall to lock some part of the strike-works. Count-wheel striking persisted in 30-hour clocks after the early eighteenth century, when the alternative system, rack-striking, was widely adopted for eight-day clocks.

Crossings
The spokes of a brass clock wheel; metal is removed from the solid wheel to reduce weight and, more importantly, the effects of inertia.

Date – day of month
An aperture below the centre of a dial can be made to show the day of the month by arranging the days on a wheel driven from the hour wheel. Date-work was often removed by early clock-repairers as a way of reducing friction losses. Modern ideas about conservation do not condone that but neither is it good conservation practice to make up replacement parts.

Dead-beat escapement
Designed by Thomas Tompion's nephew, George Graham, as an improvement on the anchor escapement; it is recognizable by the forward-pointing teeth of the escape wheel and by the precise movement of the seconds hand. Deadbeat escapements are found in accurate clocks, and clocks with a centre-mounted seconds hand, which would otherwise magnify the recoil of the hand.

Dedendum
In gear teeth, the distance of the root or base of the teeth below the pitch circle.

Detent
Latches and catches that lock and unlock the strike train.

Dial
People have faces but clocks have dials; the word derives from the same root as our words, day and diary. The dial is a plan of the day, which is divided according to the Babylonian counting system based on six and sixty, which was used by ancient Greek astronomers. The reason for sub-dividing the day into two sets of twelve hours seems to be related to the impractical aspects of trying to arrange a clock to strike from one to twenty-four (a total of three-hundred strikes) and seems to date from the late fourteenth century.

Equation of time clock
Measuring time by dividing the solar day is problematic because of annual variation, but solar time clocks are useful for astronomical observations. An equation of time clock has a built-in device that makes the correction to convert from average time to solar time.

Escapement
The escapement of a clock performs two quite distinct functions: first, it lets the power of the driving weights 'escape' at a fixed rate, turning the wheels; and, second, it gives an impulse to

the pendulum at each swing, enough to keep the pendulum swinging.

Feather
The thin strip of spring by which the pendulum hangs.

Fly
A rotating flat sheet of brass at the top of the strike train that controls the speed of the strike. It acts as a simple air-brake governor.

Gathering pallet
The single-toothed pinion that at each revolution draws the striking rack forward by one tooth.

Hands
Sometimes called fingers, point to the hours, minutes and seconds. The evolution of styles was originally documented by Herbert Cesinsky and Malcom Webster in *English Domestic Clocks* in 1913 and since then, other writers have contributed research.

Hood
The top, removable part of the case of a longcase clock. Early clocks were fitted with hoods that slide upwards to allow access to the movement, but outside museums they are almost unknown. The usual hood arrangement incorporates a horizontal runner that allows the hood to slide forwards off the case.

Hoop wheel
Usually the second wheel in a count-wheel strike train. When the count-wheel detent falls into its slot, a detent drops into a gap in the rim of the hoop wheel to lock the strike train. Hoop wheels were used in medieval clocks not only to lock the strike, but also allow the lifting piece to drop once the strike cycle commenced.

Jaws
Another word for chops (q.v.).

Lenticle
A round or oval glass window in the trunk door of a clock-case at the level of the pendulum bob. It has been suggested that the frequent use of 'bull's eye' glasses is a more recent modernization and originals were more likely to have been fitted with plain glass.

Locking wheel
Another name for count-wheel (q.v.).

Motion-work
The gearing, usually located between the dial and the movement that transfers the rotation of the going train to the hands. Most (but not all) longcase clocks have two hands and the motion-work comprises a reduction gear of 12:1, so that the hour wheel can be driven from the minute or centre wheel arbor.

Movement
The correct term for the mechanical parts of a clock – the 'works'. The movement includes the arbors, wheels and pinions, of which wheels have teeth and pinions have leaves.

Pallets
Hard steel plates of the escapement that rub against the teeth of the escape wheel.

Pinion
A small steel wheel with only a few leaves on a clock arbor, driven by the teeth of a brass wheel. In longcase clocks, steel pinions are driven by brass wheels and the only exception is in the motion-work where all the parts are brass.

Pinion of report
A pinion mounted on the end of a great wheel arbor, driving the motion-work or the count-wheel.

Pitch circle
When a pinion meshes with a gear wheel, their pitch circles just touch. It is the design datum for gear cutting.

Rack-striking
A system of counting off the hourly strikes developed by Reverend Barlow in 1670s. It is slightly more intricate that the earlier count-wheel but allows for repeating the hour without losing synchronicity.

Regulator
An accurate clock – in longcase clocks, the term is taken to mean a clock fitted with a deadbeat escapement, a maintaining-power device and a pendulum that is either inert to, or compensated for, temperature variations. In the age of enlightenment, regulators were necessary for accurate astronomy but by the latter end of the longcase period, the domestic regulator had developed as a status symbol for private houses and for clock shops.

Repeating work
In the days before convenient lighting it was possible to adapt a rack-striking clock so that, with the pull of a string, it would strike the previous hour. History does not relate how the user found the clock and its repeater cord in the dark without a light, which would have shown the time.

Roman striking
A system of striking based on having two bells of different tones to represent the Roman numerals I and V. The roman ten (X) was sounded by two strikes of the V bell. In terms of the power requirement and wheel work, it is more efficient because a twelve-hour cycle in Roman striking comprises thirty strikes, compared with seventy-eight for normal or sequential striking.

Second and a quarter pendulum
Accuracy of pendulum regulation is improved by lengthening the pendulum and a few of the early pioneer makers produced clocks with longer pendulums. Although the 1¼sec pendulum is 61in long, compared with approximately 39in for a 1sec pendulum, it can still be accommodated within a longcase of similar size. Where a subsidiary seconds dial is fitted, it may appear 'odd'. For good time-keeping, the combination of a 30-hour movement with a 1¼sec pendulum was an early success. Second and a quarter pendulums seem to have disappeared within only a few years of their introduction and consequently they are rare.

Shake
A clockmaker's word for slack or play in the mechanism. Side shake is the side-to-side movement of, for example, a pivot in its bush and end shake is axial movement of an arbor between the movement plates

Signatures
From its incorporation, the clockmaker's guild compelled members to sign their names and place of work on their clocks. Many of the early signatures were Latinized with the word 'fecit', meaning 'made it'. Signatures of commercially attractive or collectable makers are often suspect.

Six and sixty
The basis of the ancient Babylonian counting system that was used for early astronomical and astrological measurements. A circle is made up of six times 60 degrees and the four quarters of the day, sunrise to noon, noon to sunset, etc., are each divided into six divisions.

Spandrels
Fitting a round dial into a square case led to the development of corner spandrels cast in brass and often gilded. Spandrels seem to have developed in early mantle or table clocks slightly before the development of longcase clocks. Fashions in spandrel design can be a useful way of dating clocks, with the proviso that the date of a clock should be based on the estimated date of the most recent feature. With the advent of painted iron dials in the 1770s, the spandrel concept developed as painted ornamentation in the dial corners.

Time
A complicated concept to do with space and gravity – far beyond the scope of his book. At a mundane level the division of days into smaller units as a means of organizing society evolved from religious rules about prayer times.

Trunk
The part of a longcase between the base and the hood. It is fitted with a door in the front to give access to the weights and pendulum.

Bibliography

Baillie, G. H., *Watchmakers and Clockmakers of the World* (NAG Press Ltd, 1976)

Baillie, G. H., Ilbert, C. and Clutton, C. (eds), *Britten's Old Clocks and Watches* (Methuen Ltd, 1982)

Barder, Richard C. R. *English Country Grandfather Clocks* (David and Charles, 1983)

Barker, David, *The Arthur Negus Guide to English Clocks* (Hamlyn Publishing Group Ltd, 1980)

British Horological Institute (ed. Wills, Peter B.), *Conservation of Clocks and Watches* (BHI Ltd, 1995)

Bruton, Eric, *A Guide to Dating English Antique Clocks* (NAG Press Ltd, 1981)

Bruton, Eric, *Clocks and Watches* (Hamlyn Publishing Group Ltd, 1968)

Cescinsky, H. and Webster, M.R. *English Domestic Clocks* (George Routledge & Sons Ltd, 1914)

Cescinsky, H. *The Old English Master Clockmakers* (George Routledge & Sons Ltd, 1938)

Darken, Jeff and Hooper, John *English 30 Hour Clocks* (Penita Books, 1997)

De Carle, Donald, *Practical Clock Repairing* (NAG Press Ltd, 1952)

Edwardes, Ernest L., *Weight-driven Chamber Clocks of the Middle Ages and Renaissance* (John Sherratt & Son, 1965)

Edwardes, Ernest L., *The Grandfather Clock* (John Sherratt & Son, 1949)

Eldin, Herbert L., *What Wood is That?* (Viking Penguin Inc., 1969)

Ells, Anthony, *Finding and Restoring Longcase Clocks* (Crowood, 2001)

Garrard, F. G., *Clock Repairing and Making* (The Technical Press, 1948)

Hana, W. F. J., *English Lantern Clocks* (Blandford Press, 1979)

Holtzapffel, Charles, *Turning and Mechanical Manipulation* (Three Volumes) (Tee Publishing, 1993)

Law, Ivan, *Gears and Gear Cutting* (Special Interest Model Books, 1988)

Leopold, J. H., *The Almanus Manuscript* (Hutchinson & Co, 1971)

Lloyd, H. Alan *Old Clocks* (Ernest Benn Ltd, 1951)

Loomes, Brian *Grandfather Clocks and their cases* (Bracken Books, 1985)

Loomes, Brian, *The White Dial Clock* (David and Charles, 1974)

Loomes, Brian, *Country Clocks and their London origins* (David and Charles, 1976)

Loomes, Brian, *Watchmakers and Clockmakers of the World Complete 21st Century Edition* (NAG Press, 2006)

McDonald, John, *Longcase Clocks* (Country Life Books, 1982)

McQuoid, Percy, *The Age of Oak and The Age of Walnut* (Antique Collectors Club, 1987)

McQuoid, Percy, *The Age of Mahogany and The Age of Satinwood* (Antique Collectors Club, 1987)

Rees, Abraham, *Clocks, Watches and Chronometers* (David and Charles, 1970; original publication 1819)

Richards, E. G., *Mapping Time The Calendar and its History* (Oxford, 1998)

Robertson, J. Drummond, *The Evolution of Clockwork* (S. R. Publishers Ltd, 1972; original publication 1931)

Smith, Alan, *The Connoisseur Guide to Clocks and Watches* (The Connoisseur, 1975)

Taylor, V. A. & Babb, H. A., *Making and Repairing Wooden Clock Cases* (David and Charles, 1994)

Timmins, Alan, *Making an Eight day Longcase Clock* (Tee Publishing, 1981)

Ullyett, Kenneth, *In Quest of Clocks* (Spring Books, 1950)

Ullyett, Kenneth, *British Clocks and Clockmakers* (Bracken Books, 1947)

Ullyett, Kenneth, *English Clock Masterpieces* (Herbert Jenkins Ltd, 1950)

Vernon, John, *The Grandfather Clock Maintenance Manual* (David and Charles, 1983)

Weiss, Leonard, *Watchmaking in England 1760 – 1820* (Robert Hale, 1982)

Wells, P. A. and, Hooper, J., *Modern Cabinet Making* (Fox Chapel Publishing, 2006)

Index

30-hour 11, 30

addendum 186

back-cock removing 81
barley-sugar twist 179
beat-setting 94
bell-stand removing 81
birdcage 12
birdcage dismantling 88
bolt and shutter 14
brass
   casting 58
   common 57
   free-machining 57
   working 58
brazing 70
BS 978 129
bushes
   bushing machine 112
   hand replacement 110
   replacement 109

case
   backboard 33, 164
   base 34, 167
   base-board 24
   construction details 37
   evolution of 51
   feet replacement 182
   general arrangement 33
   hood rebuild 35, 177
   lacquer-work 53
   plinth replacement 182
   provincial styles 53
   seat-board 34
   trunk 34, 173

cleaning
   degreasing 89
   solutions 89
   ultrasonic 89
clicks
   faults 114
   general 30, 32
   springs 116
clock marriages 56
conservation
   definition 8
   rules 162
count wheel 45
crossing out 134
crutch repair 98
cycloidal 44, 128

dating
   collets 44
   general 40
   movement pillars 45
   wheel teeth 44
dedendum 187
depthing tool 64
dials
   early brass 47
   evolution of 46
   general 76
   painted 141
   round 49
   separating from movement 80
   white painted 48
dividing plate 134
domestic regulator 46
dry rot 157
Dutch clocks 19

eight-day 11

escapement
   anchor recoil 22
   deadbeat 17, 24
   re-facing pallets 124
   semi-deadbeat 25

false-plate 46
fly cutter 131
fly spring repair 106
French clocks 21

gear teeth uniformity 44
glue blocks 151
glues
   general 144
   making scotch glue 145
   types 146
going train
   30-hour 30, 31
   eight-day 26
gut line replacement 100
   steel 50

hinges
   hood 156
   trunk door 155
hood removing 78
Huygens endless rope 114

inspection
   annual 75
   bushes 90
   escapement pallets 91
   for wear 85
   pins 91
   pivots 90
   strike-works 91
   weekly 74

wheels and pinions   91
wheel work   86
involute gears   128

lacquer   71
long duration movements   15

maintaining power   14
marquetry   159
motion work
   bow spring   101
   eight-day   28
   removing   82
   posts   101
   testing   95
multi-tooth cutter   132

pendulum
   basic concept   9
   removing   78
   spring   32
   spring replacement   96
pin wheel faults   104
pinion replacement   137
pivots
   burnishing   108
   replacement   121
plated construction   13
posted construction   12

rack-striking   29, 45

ratchet faults   114
regulation   95
repair definition   8
repeat   29
restoration definition   8
rope replacement   100

Scandinavian clocks   21
seat boards   79
setting up   94
silvering   71
soldering   69
spandrels   47
steel
   general   58
   heat treatment   60, 61
   high carbon   59
   mild   59
   working   60
strike train
   30-hour   30, 31
   eight-day   27
strike work
   faults   103
   general   28, 29
   rack geometry   140
   rack tail   107
   recording   87
   removing   83
swan-neck moulding   184

terminology   8

tools, DIY   63

veneers
   bubbled   160
   cross-band   171
   general   61
   lines and inlays   161
   preparing   62
   removing   62
   replacing   160

warning wheel faults   104
weights
   general   32
   removing   78
wet rot   157
wheel cutting engine   70, 134
wheels
   designing   130
   make replacement   128
   mounting   137
   replace tooth   126
wood   61
wood finish
   applying shellac   149
   identifying   149
   shellac   147
   wax   147
wood mouldings   152
woodwork tools   73
woodworm   157